India in the Second Space Age of Interplanetary Connectivity

This volume discusses the emergence of space exploration as a new pivot of the global space economy in the decade of 2020s. Space exploration and human spaceflight will soon become vital strategic initiatives in the imminent Second Space Age, evolving from scientific pursuits to mega-economic projects. As the scope of international cooperation in space forays into soft science diplomacy, the Second Space Age opens opportunities for India to mount its space programme as an ambitious yet conscientious, proficient and cordial player in the global space economy.

This book,

- Explores imminent trends in space exploration and interplanetary connectivity plans, their returns to the global economy of the future and impact on the global astropolitical order.
- Analyses the techno-economic significance of India's space exploration by reviewing the legal, ethical and philosophical challenges; the limits of global space exploration policies; and the economic lacunae for the astropolitical gains.
- Examines the transformational trio of *Chandrayaan*, *Mangalyaan and Gaganyaan*; dawn of the Second Space Age; interplanetary connectivity projects; besides discussing the viability of humans becoming an interplanetary species.

Part of *The Gateway House Guide to India in the 2020s* series, this topical volume will be useful for scholars and researchers of international relations, geopolitics, foreign policy, space policy, South Asian studies, strategic studies and international trade.

Chaitanya Giri is the fellow of Space and Ocean Studies at Gateway House: Indian Council for Global Relations, India. His researches on techno-geostrategy, analyses of space industrial complex, space exploration policy, and planetary and astromaterials science. He is an affiliate scientist at the Earth-Life Science Institute at Tokyo Institute of Technology, Japan.

He has a Ph.D. in Chemistry specialising in astrochemistry space payloads from the Université Côte d'Azur, France. At the same time, he was a doctoral fellow at the Max Planck Institute for Solar System Research in Germany. He was the co-investigator of the COSAC payload on the European Space Agency's Rosetta mission to comet 67P/Churyumov-Gerasimenko. His research has earned him several fellowships and awards, including the 2014 Dieter Rampacher Prize of the Max Planck Society for the Advancement of the Science, Germany, and the 2016-2018 ELSI Origins Network Fellowship at the Tokyo Institute of Technology.

He consults India's strategic agencies and reviews India's science and technology diplomacy and delivers talks at various academic and non-academic institutions in India and abroad.

The Gateway House Guide to India in the 2020s

Series Editor: **Manjeet Kripalani**, *co-founder, Gateway House: Indian Council on Global Relations*

The Gateway House Guide to India in the 2020s explores the connections between India's globalist past to the strengths it has developed as it steps into the future, starting with the decade of the 2020s. The volumes in this series discuss a wide range of topics, which include solutions for energy independence and environmental preservation, exposition of the new frontiers in space and technology, India's trade networks, security, foreign policies and international relations. Furthermore, the series examines the embedded trading and entrepreneurial communities which are coming together to influence global agenda-setting and institution-building through platforms like the G20 and the UN Security Council, where India will take leadership roles in this decade, in the post-COVID-19 pandemic world.

This series appeals to an international audience and is directed to policymakers, think tanks, bureaucrats, professionals working in the area of politics; scholars and researchers of political science, international relations, foreign policy, world economy, politics and technology, Asian politics, South Asia studies and contemporary history; students and the general reader, seeking an understanding of what will drive India's positioning in world affairs.

India in the Second Space Age of Interplanetary Connectivity
Giri Chaitanya

Mercantile Bombay
A Journey of Trade, Finance and Enterprise
Sifra Lentin

India and the Changing Geopolitics of Oil
Amit Bhandari

For more information about this series, please visit: https://www.routledge.com/The-Gateway-House-Guide-to-India-in-the-2020s/book-series/GHGI20

India in the Second Space Age of Interplanetary Connectivity

Chaitanya Giri

Routledge
Taylor & Francis Group

LONDON AND NEW YORK

First published 2022
by Routledge
2 Park Square, Milton Park, Abingdon, Oxon OX14 4RN

and by Routledge
605 Third Avenue, New York, NY 10158

Routledge is an imprint of the Taylor & Francis Group, an informa business

© 2022 Gateway House: Indian Council on Global Relations

British Library Cataloguing-in-Publication Data
A catalogue record for this book is available from the British Library

Library of Congress Cataloging-in-Publication Data
A catalog record has been requested for this book

ISBN: 978-0-367-67851-7 (hbk)
ISBN: 978-0-367-71614-1 (pbk)
ISBN: 978-1-003-15293-4 (ebk)

DOI: 10.4324/9781003152934

Typeset in Times New Roman
by SPi Technologies India Pvt Ltd (Straive)

Dedicated to my Aai.

Contents

Illustrations

Acknowledgements

I sincerely thank my colleagues at the Indian Council for Global Relations for their continuous support and encouragement. The book was supported by various patrons of the ICGR.

Abbreviations

AECOM	Architecture Engineering Construction Operations and Management
BIMSTEC	Bay of Bengal Initiative for Multi-Sectoral Technical and Economic Cooperation
BRI	Belt and Road Initiative
BSGN	Business in Space Growth Network
BWXT	Babcock & Wilcox Technologies
CARE	Crew Module Atmospheric Re-entry Experiment
CASIC	China Aerospace Science and Industry Corporation
CASTC	China Aerospace Science and Technology Corporation
CCDeV	Commercial Crew Development Program
CEOS	Committee on Earth Observing Satellites
CGMS	Coordinating Group on Meteorological Satellites
CII	Confederation of Indian Industries
CLEP	Chinese Lunar Exploration Program
CLPS	Commercial Lunar Payload Services
CMSA	China Manned Space Agency
CNES	National Centre for Space Studies
CNSA	China National Space Administration
COMECON	Council for Mutual Economic Assistance
COSPAR	Committee on Space Research
COSTIND	Commission for Science, Technology, and Industry for National Defense
COTS	Commercial Orbital Transportation Services
COVID-19	Coronavirus Disease 2019
CRS	Commercial Resupply Services
CPC	Communist Party of China
CSIR	Council of Scientific and Industrial Research
CTBTO	Comprehensive Nuclear-Test-Ban Treaty Organization

C1XS	Chandrayaan-1 X-ray Spectrometer
DLR	German Aerospace Center
DoS	Department of Space
DRDO	Defence Research and Development Organization
DST	Department of Science and Technology
EADS	European Aeronautic Defence and Space Company
ESA	European Space Agency
ESRIC	European Space Resource Innovation Centre
ETF	Exchange Traded Fund
EVE	European Venus Explorer
FICCI	Federation of Indian Chambers of Commerce and Industry
GER	Global Exploration Roadmap
GSLV MK-III	Geosynchronous Satellite Launch Vehicle Mark III
G7	Group of Seven
G20	Group of Twenty
IAA	International Academy of Astronautics
IAF	International Astronautical Federation
ICAG-SS	Inter-Agency Consultative Group for Space Science
ICBM	Inter-Continental Ballistic Missile
IHSFP	Indian Human Space Flight Programme
IIA	Indian Institute of Astrophysics
IIT-BHU	Indian Institute of Technology Banaras Hindu University
INSARAG	International Search and Rescue Advisory Group
INTERPOL	International Criminal Police Organization
ISAS	Institute of Space and Astronautical Science
ISECG	International Space Exploration Coordination Group
ISRO	Indian Space Research Organization
JAXA	Japan Aerospace Exploration Agency
J-SPARC	JAXA Space Innovation through Partnership and Co-creation
LCROSS	Lunar Crater Observation and Sensing Satellite
MDA	Macdonald, Dettwiler and Associates Corporation
MEPAG	Mars Exploration Program Analysis Group
MIIT	Ministry of Industry and Information Technology
miniSAR	Miniature Synthetic Aperture Radar
MOM	Mars Orbiter Mission
M3	Moon Mineralogy Mapper
NASA	National Aeronautics and Space Administration
PSLV	Polar Synchronous Launch Vehicle

RADOM	Radiation Dose Monitor Experiment
RDTE	Research, Development, Testing and Evaluation
SARA	Sub-keV Atom Reflecting Analyser
SASAC	State-Owned Assets Supervision and Administration Commission of the State Council
SDG	Sustainable Development Goals
SFCG	Space Frequency Coordination Group
SIRIUS	Scientific International Research In Unique Terrestrial Station
SIR-2	Spectrometer Infrared-2
SPADEX	Space Docking Experiment
SPDR	Standard and Poor's Depository Receipt
SRE	Space Capsule Recovery Experiment
TMC	Terrain Mapping Camera
UAE	United Arab Emirates
ULV	Unified Launch Vehicle
UNCOPUOS	United Nations Committee on the Peaceful Uses of Outer Space
UNISPACE+50	United Nations Conference's Fiftieth Anniversary on the Exploration and Peaceful Uses of Outer Space
UNOOSA	United Nation's Office of Outer Space Affairs
US	United States
VEXAG	Venus Exploration Analysis Group
4G-LTE	Fourth-Generation – Long Term Evolution

1 Introduction

The 2020s is a centennial decade for the fields of modern rocketry, astronautics and space travel. The pioneering efforts of Konstantin Tsiolkovsky (1857–1935) in the Soviet Union, Hermann Oberth (1894–1989) in Germany, Robert Esnault-Pelterie (1881–1957) in France and Robert Goddard (1882–1945) in the United States (US) deserve mention.[1] Primarily unaware of each other's simultaneous endeavours, these four stirred the space programme's genesis in their home countries. Within a few years of their efforts, their innovations on reaction engines found tremendous military applications in the inter-World War era (1919–1939), first as long-range and lethal guided missiles.[2] However, while being used as missiles, these dual-use projectiles could also cross the Kármán line – the boundary between Earth's atmosphere and outer space – making them space launch vehicles.[3] The duality became apparent when the first intercontinental ballistic missile (ICBM), the Soviet Union's R-7 'Semyorka' missile, tests in May 1957, was followed by the first space launch by R-7's derivative, the Sputnik launch vehicle in October 1957.[4] The intimate link between ICBMs and space-launch vehicle technologies expanded the Cold War contest between the Soviet Union and the US in outer space and led the world into the First Space Age.[5]

After the early advances made by the US and the Soviet Union in space technologies during the intense years of the Cold War, Germany, Japan and France too began attaining top-notch space technology proficiencies beginning 1960s.[6] Their capabilities helped them make pioneering achievements in outer space. The Indian, Chinese and Japanese space programmes gathered steam after the 1980s.[7]

In any nation, a government is always the foremost benefactor to the nation's space capabilities. In that regard, national space programmes have grown to epitomise a regime or a government's futuristic ambitions and a contrivance to demonstrate world-leading human

DOI: 10.4324/9781003152934-1

competencies. However, such ambitions and demonstrations can elicit cooperation and competition between two or more goverments depending on their geopolitical dealings. All space-faring nations attempt to maximally extract geopolitical gains from cutting-edge space capabilities. They use it to set an irreversible embarkment of their footprint in outer space. Extrapolating geopolitics in outer space realms has become a new area of deliberation known as 'astropolitics.'[8]

The brief history of astropolitics, like geopolitics, which began with the advent of the First Space Age, has shown that the friendliest regimes need not collaborate all the time, as they too may compete in a non-aggressive manner. At the same time, the fiercest geopolitical opponents may not always compete but find opportunities to collaborate.

The Cold War did not prevent the US and the Soviet Union from collaborating in outer space through the Apollo-Soyuz docking mission in 1975.[9] Their civilian programmes remained competitive after that, but the détente toned down the bellicosity tremendously.[10] Russia, the successor to the Soviet Union, departed from its earlier Cold War geopolitical stance towards the US. The post-1991 Russia forfeited Soviet Union's 'Interkosmos' astropolitical bloc formed between 1978 and 1991 with the COMECON and Warsaw Pact nations; socialist nations like Cuba, Afghanistan, Vietnam and Mongolia; and non-socialist countries like India, Syria, Austria, the United Kingdom, Japan and France.[11] Russia used this forego to initiate commercial space cooperation as a significant confidence-building measure with the West, particularly the US. Since then, the International Space Station began in the late 1990s and has become a prime exemplar of cooperation between the West and Russia, with Russia singularly ferrying logistics and astronauts for nearly a decade.[12]

Today, nations with multifarious space-faring capabilities possess autonomous national space programmes, catering to civilian, military and commercial end-goals. With the severance of civilian and commercial space programmes from the military ones, the two programmes have become crucial apparatuses of diplomacy and trade. The space industry's bread and butter, i.e., satellite or spacecraft and launch vehicle design, manufacturing and services are shared by the military, civilian and commercial space programmes. Planetary exploration for long was entirely a scientific fact-finding pursuit limited to the civilian space programme and supported by academia and a few technology contractors offering tailored technologies. Few commercial stakeholders saw it as an opportunity to seek high-profile build-to-order contracts that they can boast. Nonetheless, to the uninformed citizenry, space exploration remained only an instrument of national pride.

The planetary exploration missions of the First Space Age have flown past and orbited around all planets of our Solar System at least once. Some of these missions have landed on extraterrestrial surfaces and have brought back exogeological samples from asteroids and the Moon. These missions have charted fast interplanetary transportation paths and identified preliminary solutions to low-latency and uninterrupted telecommunications, dense energy storage and atmospheric entry and re-entry during planetary space missions. There has been no crewed space mission to the Moon after the US' *Apollo* programme. However, the International Space Station has nevertheless garnered tremendous know-how of acclimatising humans for long-duration spaceflight.

The science and planning of space exploration and the technologies involved have all been an output of largely government-sponsored civilian activities. These activities have engendered a vast treasure trove of data on which the academic and industry-based specialists are training new talent with diverse expertise and experiences. However, the confines of government-sponsored laboratories can no more accommodate this vast human resource. In order to sustain this vast talent, a comprehensive commercial approach to space activities has become of essence. Tangible commercial goals will steer planetary exploration and human spaceflight activities increasingly.

Currently, the global space economy is estimated to be valued at around 360 billion dollars.[13] The highly commercial satellite industry dominates the global space economy. This approximately 271-billion-dollar industry propels the associated rocket launch, satellite manufacturing, semiconductors, electronics, energy storage, specialty chemicals and other diverse technology businesses of the diversified space industry.

The business acuity gathered by the satellite and its ancillary industries and the trend towards technology miniaturisation has motivated the space economy to move on to smaller and cost-effective satellites and spacecraft. Tandemly, this miniaturisation has also triggered the exigency for smaller, low-cost and on-demand launch vehicles, the newly begun Second Space Age's defining attribute. The same miniaturisation and its resulting cost-effectiveness is about to bring about an exponential growth in private-sector-driven commercialised planetary exploration, which this book discusses.

During the initial decades (the 1960s and 1970s) of the First Space Age, space exploration was on the science and national pride agenda only of the US and Soviet Union. The Cold War between them made it possible for large budgetary allocations to space exploration projects, particularly human spaceflight. In those decades, these were the only two nations to possess launch capacities that lifted space-craft's heavier than 2 tonnes warranted by such missions. Japan[14] and France[15] began developing mid- (1–1.5) and heavy-lift (1.5 and above) launch vehicles in the 1980s, which immediately helped them initiate their planetary exploration missions. India and China began space exploration in the early 2000s. China initiated human spaceflight in 2003 with its heavy-lift launch vehicle Long March 2F. The same rocket has now assembled its large modular space station, for executing its Chinese Lunar Exploration Program (CLEP) and its future mission to Mars.[16] India began its *Chandrayaan* lunar exploration programme with a mid-lift 1.5 tonne Polar Synchronous Launch Vehicle (PSLV).[17] It is now graduating to human spaceflight and heavy-payload planetary missions with its 2.5 tonne plus heavy-lift Geosynchronous Satellite Launch Vehicle Mark III (GSLV Mk-III) and the underdeveloped Unified Launch Vehicle (ULV).[18]

Now on the Second Space Age cusp, countries like New Zealand, Germany, the United Kingdom, which never had a rocketry portfolio in their respective space programme, have begun to harness space launch capabilities.[19] They focus on smaller and low-lift launch vehicles to cater to the growing demand for launching miniaturised satellites and spacecraft. Nations from the economically developed world like Luxembourg, Switzerland, Austria, Australia and Canada have begun to identify space applications, with Fourth Industrial Age (Industry 4.0) characteristics, to the best of their technological competencies. Luxembourg intends to explore space applications from its existing proficiencies in financial technology, banking and investments.[20] Australia offers commercial spinoffs of its rich heritage of space-based military surveillance and ground-based communications infrastructure.[21] Economically developing nations like Bangladesh, Nigeria and Malaysia have limited space competence but are making pragmatic space collaborations with nations having advanced technology competencies.

As more nations begin pursuing space activities, exploration will transform from a government-driven activity into a commercial applied-science-driven enterprise. Many nations, and the technology innovation and manufacturing ecosystems within them, will pursue space exploration clustering around a few techno-economic superpowers with whom they align politically, economically and militarily. This book deems these clusters as astropolitical blocs. Typically, astropolitical blocs would affiliate on political, economic and military aspects and share financial, industrial, natural, human and academic resources. The bloc members will seldom share these resources outside the blocs. Such clustering will be a vital facet of a geopolitically multipolar world. With more claimants to the tag of a superpower, such nations will keep a band of countries, those at lower pedestals of power than them, who could barter vital resources with them.

The Second Space Age will see many nations investing in the global space economy. They are contemplating frequent expeditionary missions to the Moon, near-Earth asteroids and Mars, some of the nearest Solar System bodies to the Earth. Astropolitical bloc leaders aim to install an interplanetary connectivity network with exclusive access only for the bloc members. These superpower countries intend to marry each of their space industrial ecosystems (consisting of academia, technology start-ups, medium-scale technology providers, large-capital technology conglomerates in collaboration with the space agencies) to construct and operate these interplanetary connectivity networks. With these entities joining in, the contest between the astropolitical blocs will be much broader in scope and consequence than the Cold War-era Space Race, which extinguished with the US reaching the Moon first. The new contest will be drawn out perhaps on the scale of centuries. It will be like the board game of *igo*, where more than two blocs will intend to secure more interplanetary and extraterrestrial expanse in their control than the other. The extant astropolitical blocs have begun to exercise their coordinated long-term plans. They are planning to set up long-term habitats – space stations, Earth's and Moon's orbits, surface crewed and robotic outposts on the Moon and Mars, and the capability of faster-crewed transportation systems on Mars.

Long-term robotic and human presence on Moon and Mars will demand continuous logistics supplies, consistent, low-latency and uninterrupted communications, low-cost and frequent transportation systems, investments in advanced materials, new energy storage technologies, long-duration life-support systems, cyber-physical systems, robots, internet-of-things, among others.[22] Many of these technologies will stem from the civilian-commercial Industry 4.0 innovation ecosystems within the astropolitical blocs. These plans are to be initiated immediately, beginning the decade of 2020s.

The techno-economic output emerging from the interplanetary connectivity networks of various astropolitical blocs will make unique flourishes to the global space economy and the larger world economy. This economic growth is a comprehensive lever on which astropolitical blocs will begin to amend to the notion of outer space as global commons, codified under the 1967 Outer Space Treaty. The amendments will delineate the interplanetary connectivity infrastructure. The blocs that have built them may not be inclined to share it with the members of opposite blocs or even nations outside the blocs altruistically. This pragmatic approach will also reflect on the agenda of various strategic multilateral groups, intergovernmental organisations and scientific and technical bodies.

The future of humanity in outer space depends on the ability of various astropolitical blocs to step forward with cooperative competition and restrain contests below conflict levels. Earth's political geography is dynamic, with nations getting dismembered, seceded and annexed every decade. Politically resilient nations will have a greater chance of deriving technological, economic, social and political advantage from the interplanetary connectivity plans. The citizens from the most resilient nations will have an opportunity to lead humanity in outer space in the coming centuries. However, they are likely to introduce their civilisational predispositions and ideological biases. A plural representation of all extinct and extant cultures is a must as humans become interplanetary species.

Notes

1 A.I. Maksimov, Founder of Cosmonautics, *Thermophysics and Aeromechanics* 14, (2007): 317–328.
2 F.H. Winter, Did the Germans Learn from Goddard? An Examination of Whether the Rocketry of R.H. Goddard Influenced German Pre World War II Missile Development, *Acta Astronautica* 127, (2016), 514–525.
3 M. Craven, 'Other Spaces': Constructing the Legal Architecture of a Cold War Commons and the Scientific-Technical Imaginary Outer Space. *European Journal of International Law* 30, (2019), 547–572.
4 C. Lardier & S. Barensky, Designing the Semyorka. In: *The Soyuz Launch Vehicle*. Springer Praxis Books. Springer, New York, NY (2013). https://doi.org/10.1007/978-1-4614-5459-5_2
5 A. Gorman & B. O'Leary, An Ideological Vacuum: The Cold War in Outer Space. In: *A Fearsome Heritage: Diverse Legacies of the Cold War*, Ed. J. Schofield & W. Cocroft. Routledge, London and New York (2007), ISBN 9781598742596.
6 A. Dupas, Asia in space: The Awakening of China and Japan, *Space Policy* 4, (1988), 31–40.
7 A. Lele, Asian Space Race: Rhetoric or Reality? *Springer India* (2013), ISBN 9788132207320.
8 E.C. Dolman, Geostrategy in the Space Age: An Astropolitical Analysis, *Journal of Strategic Studies* 22, (1999), 83–106.
9 O. Krasnyak, The Apollo-Soyuz Test Project: Construction of an Ideal Type of Science Diplomacy, *The Hague Journal of Diplomacy* 13, (2018), 410–431.

10 T. Ellis, "Howdy partner!"- Space Brotherhood, Detente and the Symbolism of the 1975 Apollo-Soyuz Test Project, *Journal of American Studies* 53, (2019), 744–769.

11 C. Burgess & B. Vis, Interkosmos: The Eastern Bloc's Early Space Program, *Springer-Praxis* (2016), ISBN 9783319241616.

12 W.H. Lambright, Administrative Leadership and Long-Term Technology: NASA and the International Space Station, *Space Policy* 47, (2019), 85–93.

13 K.W. Crane, E. Linck, B. Lal, & R.Y. Wei, Measuring the Space Economy: Estimating the Value of Economic Activities in and for Space, *IDA – Science & Technology Policy Institute*, (2020), https://www.ida.org/-/media/feature/publications/m/me/measuring-the-space-economy-estimating-the-value-of-economic-activities-in-and-for-space/d-10814.ashx

14 T. Tadakawa, Japan's Launch Vehicle Program Update, *SAE Transactions* 96, (1987), 219–228.

15 R. Deschamps, Ariane in the World Launch Vehicle Market, *Space Policy* 1, (1985), 76–81.

16 N. Goswami, China in Space: Ambitions and Possible Conflicts, *Strategic Studies Quarterly* 12, (2018), 74–97.

17 J.N. Goswami & M. Annadurai, Chandrayaan-1: India's first planetary science mission, *Current Science* 96, (2009), 486–491.

18 S. Somnath, S. Unnikrishnan Nair, T. Sivamurugan, & S.L.N. Desikan, Flight Performance of Crew Escape System during Pad Abort Condition, *Current Science* 120, (2021), 81–88.

19 C. Giri, A Space Exploration Industry Agenda for India, *Gateway House: Indian Council for Global Relations*, Paper No. 23, 2020, https://www.gatewayhouse.in/wp-content/uploads/2020/05/26-May_A-space-exploration-industry-agenda-for-India_Chaitanya-Giri_Final.pdf

20 L. Thailly & F. Schneider, Luxembourg set to Become Europe's Commercial Space Exploration Hub with New Space Law, *Ogier*, August 2017, https://www.ogier.com/news/the-luxembourg-space-law

21 W. Flentje, S.E. Pearce, K.C. Clayfield, A.A. Held, & P. Crosby, A Roadmap for Space Industry Development Through Public-Private Collaboration in Australia. In: *31st IAA Symposium on Space Policy, Regulations and Economics, 69th International Astronautical Congress*, 1 October 2018, Bremen, Germany.

22 P. Haschemi & S. Khodabakshi, Industry 4.0 Will Revolutionize the Space Market, *Room – Space Journal of Asgardia* 18, (2018), https://room.eu.com/article/industry-40-will-revolutionise-the-space-market

2 India and a brief history of time in space exploration

The summer of 2020 has been a watershed season for the Indian space programme. Fifty years since the formal inception of the Indian Space Research Organisation (ISRO), in 1969,[1] New Delhi undertook far-reaching reforms in its approach and perception of outer space activities for the first time. The foremost of the several pathbreaking space policy reforms involves giving India's private sector a pedestal equal to the state-run entities in India's various space activities.[2]

Presently India has a sparsely populated private space ecosystem, limited to a few start-ups, micro, small and medium enterprises, large technology companies and multi-sectoral conglomerates that have operated time and again as vendors to ISRO and its sister Department of Space (DoS) laboratories. Through the space reforms, the Indian government has begun to recognise the indispensability of the private space ecosystem in ushering India into the Second Space Age. It has now pledged to encourage private sector's participation in India's future space projects and missions, assist them in plugging into the space projects with India's strategic partners and conceptualising and executing innovative space missions indigenously. By doing so, the Indian private sector can become a leading element of India's civilian and commercial planetary exploration missions.

During the First Space Age, the Indian space programme focused on Earth-centric applications and neglected planetary exploration as a fanciful undertaking.[3] New Delhi was slow and modest about its space programme, and the private sector's participation was confined to small contractual offerings giving them no scope for innovation. On the contrary, many Western nations cultivated space competencies within their military industrial complexes and more so under the wings of their large private defence contractors like Boeing, EADS (now Airbus Defence and Space), Lockheed Martin, Thales and Northrop Grumman. These defence contractors raised space contractual

DOI: 10.4324/9781003152934-2

manufacturing and innovation sections in their establishments. Over the years, these space wings of defence contractors have carried out research, development, testing and evaluation (RDTE) of space launch systems, spacecraft, satellites, space station, landers, rovers, payloads and astronaut life-support systems into their portfolio. This well-planned assimilation of the private sector helped Western nations acquire an edge over the state-run space programme of the Soviet Union and its COMECON partners.

India, with its space sector reforms, is aiming to mature its space innovation and industrial ecosystems to match to the demands of the Second Space Age. The business models demonstrated by new-age space companies from the democratic world are construed by the Indian ecosystems as exemplars to set up similarly or even better equipped domestic companies and models. The Indian space ecosystem aims to seek value from ISRO's technical and commercial prowess and graduate it to higher echelons. Through these space reforms, the Indian Prime Minister's Office intends to strengthen the Indian space programme by galvanising the voluminous human, technical and monetary resources that the Indian private sector could accommodate and plug them to participate in futuristic space activities. The excellent appraisals garnered by ISRO's select planetary exploration, space-based astronomy and the recently initiated human-spaceflight missions have prompted New Delhi to begin nurturing the Indian private sector to commercialise space technologies.

New Delhi has assessed the empathetic public sentiment in favour of planetary exploration, as has been witnessed during the *Chandrayaan* (moon-bound), *Mangalyaan* (Mars-bound) and *Gaganyaan* (human spaceflight) series of missions. The supportive opinion has been of utmost importance in India as its polity for long viewed scientific exploration as a fantastic but socio-economically irrelevant space activity.[4] The swelling democratisation of global space activities has begun to erode such Cold War-era socialist slants of India's polity.

The Armada approach of new entrants in space exploration

The Cold War race to the Moon between the US and the Soviet Union ended when the US successfully landed the first astronauts on *Mare Tranquillitatis* in 1969 with the Apollo 11 mission.[5] Later in the 1970s, both the US and the Soviet Union peacefully sustained their respective *Apollo* and *Luna* series of lunar orbiting, landing and fly-by missions.[6]

However, by the middle of the 1970s, the US and Russian lunar exploration programmes suffered from fatigue and became bereft of any motivation.[7] The entire decade of the 1980s went away without a mission to the Moon. However, during this same period, many new countries entered the arena of planetary exploration.

In its early years, the Japanese space programme focused on launching communication and meteorological satellites using its earliest *Nippon-1* and *Nippon-2* launch vehicles. The US' *Thor* missile-turned-launch vehicle was used to build these two launch vehicles. Mitsubishi Heavy Industries made the second stage, and the third stage was made using *Castor* solid rocket motor provided by Thiokol, a subsidiary of Lockheed Martin.[8] Taking lessons from *Nippon-1* and *Nippon-2* launchers, the Japanese Institute of Space and Astronautical Science (ISAS), which later merged into Japan Aerospace Exploration Agency (JAXA) in 2003, invested in building its indigenous *Mu* series of launch vehicles. Between 1981 and 1983, the *Mu* rockets successfully launched several scientific 'astronomy' spacecrafts weighing up to 300 kg in the low-Earth orbit (180 km from sea level) to medium-Earth orbit (up to 2000 km from sea level).[9]

During this phase, not only Japan but Western Europe, particularly the European Space Agency (ESA) and the French space agency, CNES, were ready with their respective proven *Mu* and *Ariane* series of launch vehicles. Their readiness prompted them to initiate their first interplanetary missions. However, their target was not a planet or the Moon but the Halley's Comet, making an apparition in 1986.[10] JAXA's now-constituent ISAS was preparing the *Suisei* and *Sakigake* missions while the ESA was readying the *Giotto* mission. The US redirected its *Pioneer 7*, *Pioneer Venus Orbiter* and the *International Sun-Earth Explorer*. The Soviet Union too redirected its *Vega-1* and *Vega-2* spacecraft to approach Halley's Comet and make fly-by rendezvous-based scientific investigations.

Since all these spacecrafts were bound for Halley's Comet, it made sense that space agencies working on these spacecrafts cooperate with each other and coordinate the technical specifications of their respective spacecraft and payloads. This rationale resulted in the formation of an Inter-Agency Consultative Group for Space Science (ICAG-SS) in September 1981, with the first members being NASA (US), Rosaviakosmos (Soviet Union), ISAS (Japan) and the ESA (Europe).[11] The multilateral space diplomatic efforts of ICAG-SS were a thumping success. The flotilla of spacecraft that made vital scientific investigations of Halley's Comet caught the label *Halley's Armada*.

In 1990, Japan became the third country after the US and the Soviet Union to orbit the Moon with its *Hiten* spacecraft.[12] Apart from payloads from Japanese laboratories, ISAS continued with interplanetary diplomacy set by ICAG-SS and was able to invite a payload jointly developed with Technische Universität München from Germany for *Hiten*.[13] In 1992, ISAS collaborated with NASA and was again lunar-bound with an electromagnetic measurements mission to the Moon known as *Geotail*.[14] Later in 1994, the US' Ballistic Missile Defense Organization (today's Missile Defense Agency) and NASA jointly returned to the Moon with the Clementine mission. This multi-instrument mission consisted of near-infrared cameras, LIDAR systems and high-resolution cameras,[15] which led to the first clues of water ice at the lunar south pole. This discovery motivated many other nations to join in the exploration of the Moon.

In 1996, in the backdrop of the successful *Hiten* and *Clementine* lunar missions, the Chinese Academy of Sciences (CAS), the China National Space Administration (CNSA) and the Commission for Science, Technology and Industry for National Defense (COSTIND) began giving due consideration to lunar exploration. By 1996, the CNSA and its allied launch vehicle R&D agency, the China Aerospace Science and Technology Corporation (CASTC), had developed a proven launch vehicle that could lift around 2.5 tonnes in geostationary transfer orbit and thus in lunar-transfer orbit.[16]

The CNES and the ESA too, around the middle of the 1990s, were in the advanced stages of building the heavy-lift vehicle Ariane 5[17]; the ESA simultaneously was also amassing competence in building and operating interplanetary probes. The ESA's interplanetary missions included *Huygens* lander bound for Saturn's natural satellite Titan launched in 1997 as part of the NASA-ESA Cassini-Huygens mission[18] and the ESA-led Mars Express orbiter mission launched in 2003.[19]

By the late 1990s, India successfully and consistently demonstrated medium-lift launch capability, of up to 1.5 tonnes to GTO, with its indigenously built PSLV.[20] This propensity paved the way for the Indian Academy of Sciences and the Astronautical Society of India to conceptualise a lunar mission and form a Moon Mission Task Force in the years 1999 and 2000, respectively.[21] On 15 August 2003, Prime Minister Atal Bihari Vajpayee announced the launch of *Chandrayaan-1*, India's first space exploration mission, by the year 2008.[22]

Between 2003 and 2010, the period again witnessed an armada of spacecraft bound for the Moon that included the ESA's *SMART-1*, ISRO's *Chandrayaan-1*, CNSA' *Chang'e-1*, US' *Lunar Reconnaissance*

Orbiter-LCROSS and JAXA's *Kaguya*.[23] There was tremendous cross-pollination of payloads, data sharing, telemetry, tracking and deep space communication support between most space agencies, which only signified space diplomacy's workability via both cooperation and coordination.

Chandrayaan-1 has been one of the world's most significant space diplomacy efforts, not because ISRO chose to affiliate with expert space agencies and laboratories. However, it chose to accommodate agencies and laboratories with less experience in lunar science but had the scientific and technical acumen to develop payloads. One such team was from the Bulgarian Academy of Sciences, which provided its instrument RADOM-7 radiation dosimeter for *Chandrayaan-1*. The earlier versions of RADOM-7 were flown twice on the International Space Station, once via the US Destiny module in 2001 and later by the European Columbus module in February 2008. It also flew on the experimental flights of Russia's FOTON M2 and M3 experimental space capsules.[24] These RADOM-7 predecessors were operating in the low-Earth orbit, but with *Chandrayaan-1*, the Bulgarian Academy of Sciences was able to undertake studies during the Earth-to-Moon flight as in the Moon's orbit.

Chandrayaan-1 also flew the Spectrometer Infrared-2 (SIR-2) that was built by the German space science institute the Max Planck Institute for Solar System Research, the ESA with support from the Polish Academy of Sciences (Poland) and the University of Bergen (Norway). The spacecraft also piggybacked a British payload built jointly by the ESA and the Rutherford Appleton Laboratory, the *Chandrayaan-1* X-ray Spectrometer (C1XS) as well as a pan-European ESA payload known as Sub-keV Atom Reflecting Analyser (SARA).[25] ISRO also sought support from the technically and operationally seasoned space agencies for *Chandrayaan-1*. The US provided time slots with the three radio antennae of its Deep Space Communications Network based in Canberra, Madrid and California.[26] The two US payloads on *Chandrayaan-1* came from both civilian and military quarters. The Brown University and NASA's Jet Propulsion Laboratory developed the Moon Mineralogy Mapper (M3). Whereas the dual-use technology developers of the Applied Physics Laboratory at Johns Hopkins University and the Naval Air Warfare Center developed the Miniature Synthetic Aperture Radar (miniSAR).[27]

Chandrayaan-1 led to the pathbreaking discovery of water molecules both with the US payload M3 and the ISRO-built Moon Impactor Probe. SARA deduced the interaction between the oxygen present in the regolith on the lunar surface and the hydrogen emanating from the solar winds to form hydroxyl radical, a precursor to water.[28] ISRO's Terrain Mapping Camera (TMC) was pivotal in identifying lava tubes and caves on the surface of the Moon.[29] These and many other studies, all of *Chandrayaan-1*, assisted in multinational lunar diplomacy. Indian scientists were able to generate numerous scientific papers with their overseas counterparts. They participated in numerous European and US missions as co-investigators, young Indian doctoral and postdoctoral students availed opportunities to train in overseas laboratories, which all went on to strengthen track II diplomatic interactions.

The subsequent *Chandrayaan-2*, in the early days of mission design, was a joint India–Russia mission with multinational payloads.[30] In 2007, ISRO and the Russian space agency, Roscosmos, were contemplating a *Chandrayaan-2* with ISRO tasked to build and operate the orbiter and Russia tasked to build and operate a lander and rover. After procrastinating for six years, both the agencies called off their agreement in 2013.[31] India later on pursued *Chandrayaan-2* singularly in 2019.

India in the US-dominated domain of Mars exploration

During most years of the 1990s and 2000s, Indian scientists, particularly those at the Physical Research Laboratory, Ahmedabad, and National Geophysical Research Institute, Hyderabad, were undertaking preliminary research on the upper atmosphere of Mars and Martian meteorites. During this phase of early-stage Mars research in India, the DoS was also taking cues from various orbiter missions bound for Mars.

In 1989, Martian research saw an upsurge after a long lull period of nearly 15 years after NASA's *Viking* programme and the Soviet Union's Mars programme. The surge was due to NASA's establishment of the Mars Exploration Program Analysis Group (MEPAG), a study group aiming to provide scientific inputs for the study of Mars.[32] The MEPAG sprang into action within six years, with its first mission Mars *Global Surveyor* followed by other orbiters, lander and rover missions launched at regular intervals of 3–4 years and occasionally with international partners (Table 2.1). These missions investigated the Martian surface and subsurface geology, its tenuous atmosphere, habitable zones and biochemical signatures from any extinct or extant life forms on Mars.

The DoS, the prime Indian government body leading ISRO and other space laboratories, undertook feasibility studies for a mission

Table 2.1 Scientific goals of orbiters that visited Mars between 1990 and 2010

Mission	Agency	Launch Date	Termination Date	Scientific Goals
Mars Observer	NASA	25 September 1992	21 August 1993	Determine elemental geology and mineralogy; define gravitational fields; ascertain nature of Martian magnetic fields; determine abundance, cycle, source and sink of liquids and gases; comprehend the structure and atmosphere of Martian atmosphere
Mars Global Surveyor	NASA	7 November 1996	2 November 2006	Study composition and distribution of surface minerals, rocks and ice; monitor global weather on Mars; study seasonal variations in polar ice, atmospheric dust and clouds; characterise surface geological features and processes; determine the nature of magnetic field around Mars; identify landing sites of interest for next-generation landers and rovers
Nozomi	JAXA	4 July 1998	31 December 2003	Study of escape of Mars atmosphere; study of interaction of upper atmosphere with solar winds; evaluate structure of Martian ionosphere; penetration of solar wind in Martian atmosphere

Mars Climate Orbiter	NASA	11 December 1998	23 September 1999	Identify signatures of climate change in Martian past; monitor water vapour and dust in Martian atmosphere; record temperature profile of the atmosphere
2001 Mars Odyssey	NASA	7 April 2001	ongoing	Map the global distribution of subsurface water on Mars; infrared mapping of Mars surface; measure radiation environment on Mars
Mars Express	ESA	2 June 2003	ongoing	Detect polar ice caps and subsurface water on Mars; determine elemental composition of atmosphere; quantify atmospheric temperature and pressure; investigate relations between Martian upper atmosphere and solar winds; determine mineral chemical composition of the Martian surface
Mars Reconnaissance Orbiter	NASA	12 August 2005	ongoing	Observe atmospheric circulation and seasonal variations on Mars; locate and characterise geological processes altering Mars surface; identify signs of past and present hydrological activities on Mars surface

bound for Mars, the next closest destination to the Earth, after the success of *Chandrayaan-1*. However, unlike the other Mars exploring nations, India's Mars mission was to happen with minimum innovative contributions from the industry and inputs only from research groups residing in DoS laboratories.

Prime Minister Manmohan Singh and his United Progressive Alliance government announced the Mars Orbiter Mission (MOM) in August 2012 for launch within a short span of slightly more than a year after the announcement.[33] This short period disallowed DoS from inviting laboratories from partner countries to share their payloads. It was also not entirely possible for DoS to get the much necessary space-proven tag for the heavy-lift GSLV Mk-III during this period as the launch vehicle had experienced launch failure in 2010.[34] Had GSLV Mk-III been ready, MOM could have been much larger in spacecraft mass and the variety of scientific payloads. With the PSLVs lighter lift capability, MOM's payload suite was limited to only five low-mass scientific payloads and becoming the lightest payload-carrying spacecraft to fly to Mars[35] (Table 2.2). A lighter payload mass resulted in a considerable reduction in the instruments' technical complexity, which reduced the scientific output from MOM well before the mission began. Therefore, New Delhi termed MOM a mere 'technology demonstrator' whose principal goals were only to securely travel through

Table 2.2 Physical mass of prominent Mars-bound spacecraft

Spacecraft	Spacecraft Dry Mass (kg)	Spacecraft Wet Mass (kg)	Payload Mass (kg)
Mars Observer	1124.5	1440.5	156.6[40]
Mars Global Surveyor	767[41]	1060[42]	76.3[43]
Nozomi	258	540	30[44]
Mars Climate Orbiter	338	629	42[45]
2001 Mars Odyssey	331.8	680.5	44.5[46]
Mars Express	796	1223	187[47]
Mars Reconnaissance Orbiter	892	2180	139[48]
Mars Orbiter Mission	500	1350	14

the trans-planetary path, attain the elliptical orbit and accomplish few rudimentary scientific investigations.[36] Since the time window from announcement to launch was short, and the payload capacity was smaller than usual, MOM's payloads largely came out of DoS laboratories, and most were a heritage from *Chandrayaan-1*. The Methane Sensor on Mars,[37] came from ISRO's remote sensing heritage. Nearly seven years since its launch, the MOM is still operational, in 2021, in the Martian elliptical orbit and has been intermittently delivering scientific data.

Although the MOM spacecraft and payload are of indigenous origin, India depended on NASA's global deep space communications network support. The basis of this support was the US-India Joint Working Group on Civil Space Cooperation, first initiated by the US Department of State and Government of India's Ministry of External Affairs in March 2005.[38] After the successful orbital insertion of MOM into the Mars orbit, ISRO formed a vital Joint Mars Working Group with NASA in 2014. This working group's formation sets a platform for ISRO and NASA to collaborate, share information and provide necessary technical support whenever each of them undertakes a Mars mission.[39] *Chandrayaan-1* and MOM led the foundation of India's planetary exploration programme. They initiated a robust diplomatic engagement with the US, which leads in expertise, experience and vision necessary for the future of lunar and Martian exploration.

India's plans for the neglected Venus

In 2019, ISRO released an announcement of an opportunity for payloads to go on a Venus orbiter, *Shukrayaan-1*, for launch around 2023.[49] This mission has its antecedents in the intense advocacy campaigns led by the NASA's Venus Exploration Analysis Group (VEXAG) that consisted of international scientists from Europe working on the ESA's *Venus Express* mission and Japan on the *Akatsuki* mission.[50] For 30 years, the US was unable to execute a Venusian mission, as lunar, Martian and outer solar system exploration was always high on priority. This neglect prompted the VEXAG community, which came about in 2005, to begin plugging into Venusian missions carried out by certain agencies and promoting other space agencies' need to explore Venus. In 2007, the VEXAG enthused European scientists to work on an in-situ (sampling) mission concept known as the European Venus Explorer (EVE). The EVE could have been part of the ESA's Cosmic Vision 2015–2025 programme that included an orbiter and a balloon payload.[51] However, even in 2021, the ESA has not yet chosen the EVE concept.

Venus poses extreme challenges in terms of orbital remote sensing and *in situ* sampling because of its dense atmosphere and terrain that is detrimental to electronics and mechanical equipment. It has not been a neglected planet, but it has always demanded exclusive mission design much different from those used for lunar or Martian missions. Moreover, Venus's exploration needed a robust scientific rationale so that it could take away some attention from Moon or Mars. The detection of phosphine in the clouds of Venus in 2020 has created a strong buzz to revisit the planet as vociferously as space agencies are interested in Mars.[52]

ISRO scientists began participating in the global VEXAG community in 2012.[53] As the first outcome of this participation, ISRO entered into an understanding with JAXA to downlink and jointly studied data on Venus' atmosphere generated by the latter's *Akatsuki* mission.[54] ISRO intends to collaborate with the CNES on *Shukrayaan-1*, where French scientists have offered a payload called Venus Infrared Atmospheric Gases Linker, which they have co-developed with Roscosmos. French scientists have also proposed a balloon payload, the earlier part of the EVE concept, to piggyback on *Shukrayaan-1*.[55] India's *Shukrayaan-1* will benefit from collaborations with the two Venus-exploring groupings, one led by VEXAG and has the ESA and JAXA contributions, whereas the other led by French–Russian collaborations. To that end, *Shukrayaan-1* can become a mission bridging the scientific objectives of NASA's proposed DAVINCI mission[56] which is yet awaiting its launch announcement, and Roscosmos' *Venera-D*, which may see a 2026 launch.[57]

Settling India in low-Earth orbit with *Gaganyaan*

Prime Minister Narendra Modi announced the Indian Human Spaceflight Programme (IHSFP) on 15 August 2018, keeping with the tradition of announcing ambitious space missions on India's Independence Day.[58] The IHSFP was, however, in the works for a long time. The serious deliberation on human spaceflight began in 2006 during a meeting organised at the Antariksh Bhavan in Bengaluru.[59] The ISRO had then contemplated using the GSLV MK-II or GSLV MK-III launch vehicle, but neither of the two had acquired space-proven credentials then.

The Space Capsule Recovery Experiment (SRE-1), launched on the PSLV in 2007, for the first time, gave ISRO the opportunity to test a space capsule's in-orbit operations, its onboard payload functioning, high-temperature and high-speed atmospheric re-entry and recovery after touch-down in the Bay of Bengal. Through these SRE-1 tests, ISRO was able to undertake RDTE of thermally insulated aerody-namic structural materials and structure of the capsule, the naviga-tion, guidance and control during in-orbit operations and return, and its deceleration and launch systems.[60] ISRO worked for a few years on the SRE-II until the project was put to rest indefinitely. Both SRE-1 and SRE-II were supposed to add to the knowledge of in-orbit pay-load management and detailed scientific investigations. SRE-II was supposed to undertake biological experiments with payloads delivered by the Centre for Cellular and Molecular Biology of Hyderabad and scientists from JAXA.[61] The SRE-II would have operated on the lines of the ESA's BIOPAN and Roscosmos' FOTON series of biological experimental space capsules. However, ISRO has not classically expressed interest in carrying out extensive research in space biological studies.

The United Progressive Alliance's second term emphasised MOM, but it still could not commit long-duration funding for the human space flight programme. There were primarily three reasons. The tech-nical reason was that the 3-tonne plus GSLV MK-III did not acquire the essential space-proven credentials during their tenure; the second reason was the then underperforming national economy. The third reason was the anti-incumbency for the then ruling dispensation.

The earliest sign of Prime Minister Narendra Modi's government's support for human spaceflight appeared as it approved the Crew Module Atmospheric Re-entry Experiment (CARE) launch of 3.7-tonne boilerplate space capsule in December 2014 on the maiden GSLV MK-III launch.[62] In July 2018, ISRO undertook CARE's launch-escape test from the launch pad, and with this, the CARE boil-erplate capsule began its march towards qualifications for crewed orbital tests.[63] Within a month of the launch escape test, Prime Minister Narendra Modi announced the IHSFP.

The announcement of the IHSFP attracted immediate support from France and Russia. The CNES agreed to share its know-how on space medicine.[64] Roscosmos, since then, has been training Indian astronauts at the Yuri A. Gagarin State Scientific Research-and-Testing Cosmonaut Training Center.[65] The Russian assistance to the IHSFP has antecedents from the Soviet-era Interkosmos programme when both French and Indian cosmonauts had participated. During the Interkosmos era, the Indian Air Force was the participating agency from the Indian side. This time DoS has joined hands with the Indian

Air Force, the Indian Navy, the Department of Science and Technology (DST), the Council of Scientific and Industrial Research (CSIR), the Defence Research and Development Organisation (DRDO) and Hindustan Aeronautics Limited. Therefore, the IHSFP has become a pan-governmental undertaking.[66]

The first crewed orbital IHSFP mission will take place with a highly upgraded version of the CARE boilerplate known as the *Gaganyaan*. ISRO is also working towards a separate low-Earth orbit rendezvous, docking and robotic remote arm experiment known as Space Docking Experiment (SPADEX), the precursor to India's modular space station that will go in orbit by the early 2030s.[67] Until then, the DoS could identify and absorb the best practices from the US, Russian, European, Chinese and Japanese human spaceflight projects. The DoS will be carefully monitoring the development of next-generation space stations called Future Platforms[68] by the International Space Exploration Coordination Group (ISECG). India's space station programme will likely benefit from the emphasis on public–private synergies as mandated by the space reforms of May 2020.

India collaborates evenly with all major geopolitical clusters, which signifies space diplomacy success and represents optimal functioning of its doctrine of strategic autonomy even in outer space. India realises its inadequacies with the rapidly evolving commercially driven global global space economy. The reforms of May 2020 are an attempt to fill the glaring gaps, a necessary course correction before entering the deciding Second Space Age.

Notes

1 "About ISRO," Indian Space Research Organization, https://www.isro.gov.in/about-isro
2 "Historic reforms initiated in the Space sector: Private sector participation in Space activities approved," Prime Minister's Office – Press Information Bureau, Government of India, accessed 24 June 2020, https://pib.gov.in/PressReleasePage.aspx?PRID=1633892
3 R. Aravamudan, "Vikram Sarabhai: His vision of India as a space power and its fulfilment," *Current Science* 118, (2020), 1199–1202.
4 K. Kasturirangan & M.D. Joglekar, "Social dimensions of India's space programme," *Current Science* 108, (2015), 310–312.
5 R.D. Launius, "First Moon landing was nearly a US-Soviet mission," *Nature* 571, (2019), 167–168.
6 M.S. Robinson, J.B. Plescia, B.L. Jolliff, S.J. Lawrence, "Soviet Lunar sample return missions: Landing site identification and geologic context," *Planetary and Space Science* 69, (2012), 76–88.
7 R.D. Launius, "Planning the post-Apollo space program: Are there lessons for the present?" *Space Policy* 28, (2012), 38–44.

8 A. Kayama & Y. Takenaka, "The second stage propulsion system for N-launch vehicle," *Acta Astronautica* 7, (1980), 753–771.

9 Y. Matogawa, "Lessons from half a century experience of Japanese solid rocketry since Pencil rocket," *Acta Astronautica* 61, (2007), 1107–1115.

10 C. Stelzried, L. Efron, J. Ellis, "Halley Comet missions," Accessed from the *Jet Propulsion Laboratory* website, https://ipnpr.jpl.nasa.gov/progress_report/42-87/87X.PDF

11 D.N. Baker, "The Inter-Agency Consultative Group science campaigns," *Physics and Chemisty of the Earth, Part C: Solar, Terrestrial & Planetary Science* 24, (1999), 29–36.

12 K. Usegi, H. Matsuo, J. Kawaguchi, & T. Hayashi, "Japanese first double lunar swingby mission "Hiten"," *Acta Astronautica* 25, (1991), 347–355.

13 H. Iglseder, R. Münzenmayer, H. Svedhem, & E. Grün, "Cosmic dust and space debris measurements with the Munich dust counter on board the satellites Hiten and BremSat," *Advances in Space Research* 13, (1993), 129–132.

14 A. Nishida, "The Geotail Mission," *Geophysical Research Letters* 21, (1994), 2871–2873.

15 T.C. Sorensen, P.D. Spudis, "The Clementine mission – A 10-year perspective," *Journal of Earth System Science* 114, (2005), 645–668.

16 M. Aliberti, China's way to the Moon. In: *When China Goes to the Moon…. Studies in Space Policy*, Vol. 11, Springer, Cham. (2015) https://doi.org/10.1007/978-3-319-19473-8_4

17 L. Ravet, "The ARIANE 5 launcher improvements," *Air & Space Europe* 2, (2000), 68–72.

18 T. Owen, "Huygens rediscovers Titan," *Nature* 438, (2005), 756–757.

19 R. Schmidt, "Mars Express – ESA's first mission to planet Mars," *Acta Astronautica* 52, (2003), 197–202.

20 V. Gowarikar & B.N. Suresh, "History of rocketry in India," *Acta Astronautica* 65, (2009), 1515–1519.

21 N. Bhandari, "Chandrayaan-1: Science goals," *Journal of Earth System Science* 114, (2005), 701–709.

22 "ISRO Scientists call on PM," Prime Minister's Office – Press Information Bureau, Government of India, accessed 7 November 2003, https://archive.pib.gov.in/archive/releases98/lyr2003/rnov2003/07112003/r071120038.html

23 C. Heinicke & B. Foing, "Human habitats: Prospects for infrastructure supporting astronomy from the Moon," *Philosophical Transactions of the Royal Society A: Mathematics, Physical and Engineering Sciences* 379, (2021), 2188, https://doi.org/10.1098/rsta.2019.0568

24 T.P. Dachev, B.T. Tomov, Yu.N. Matviichuk, P.S. Dimitrov, S.V. Vadawale, J.N. Goswami, G. De Angelis, & V. Girish, "An overview of RADOM results for earth and moon radiation environment on Chandrayaan-1 satellite," *Advances in Space Research* 48, (2011), 779–791.

25 J.N. Goswami & M. Annadurai, "Chandrayaan-1 mission to the Moon," *Acta Astronautica* 63, (2008), 1215–1220.

26 N. Vighnesam, A. Sonney & N.S. Gopinath, "India's first lunar mission Chandrayaan-1 initial phase orbit determination," *Acta Astronautica* 67, (2010), 784–792.

27 S. Nozette, P. Spudis, B. Bussey, R. Jensen, K. Raney, H. Winters, C.L. Lichtenberg, W. Marinelli, J. Crusan, M. Gates, & M. Robinson, "The

Lunar Reconnaissance Orbiter Miniature Radio Frequency (Mini-RF) Technology Demonstration," *Space Science Reviews* 150, (2010), 285–302.

28 R. Sridharan, S.M. Ahmed, T.P. Das, P. Sreelatha, P. Pradeepkumar, N. Naik, & G. Supriya, "'Direct' evidence for water (H$_2$O) in the sunlit lunar ambience from CHACE on MIP of Chandrayaan I," *Planetary and Space Science* 58, (2010), 947–950.

29 A.S. Arya, R.P. Rajasekhar, G. Thangjam, Ajai, & A.S. Kiran Kumar, "Detection of potential site for future human habitability on the Moon using Chandrayaan-1 data," *Current Science* 100, (2011), 524–529.

30 C. Mathieu, "Assessing Russia's space cooperation with China and India – Opportunities and challenges for Europe," *Acta Astronautica* 66, (2010), 355–361.

31 "Chandrayaan-2," *Department of Space – Press Information Bureau, Government of India*, Accessed on 14 August 2013, https://pib.gov.in/newsite/PrintRelease.aspx?relid=98239

32 W.H. Lambright, *Why Mars? NASA and the Politics of Space Exploration*, Baltimore: John Hopkins University Press, 2014.

33 P. Bagla, "Qualms About India's Plan for a 2013 Mars Mission," *Science*, Accessed on 21 August 2012, https://www.sciencemag.org/news/2012/08/qualms-about-indias-plan-2013-mars-mission

34 "GSLV-F06 Failure-Preliminary findings and Further steps," Indian Space Research Organization, Accessed on 31 December 2010, https://www.isro.gov.in/update/31-dec-2010/gslv-f06-failure-preliminary-findings-and-further-steps

35 A. Bhardwaj, "Indian Mars Orbiter Mission," *40th COSPAR Scientific Assembly*. Held 2–10 August 2014, in Moscow, Russia, Abstract id. C3.2-14-14.

36 S. Kumar, "India joins elite Mars club," *Nature*, 24 September 2014, https://www.nature.com/news/india-joins-elite-mars-club-1.15997

37 K. Mathew, S.S. Sarkar, A.R. Srinivas, M. Dutta, M. Rohit, H. Seth, R. Kumaran, K. Pandya, A. Kumar, J. Sharma, J. Desai, A. Patel, V. Patel, P. Shukla, S. Manthira Moorthi, A.K. Singh, A. Gupta, J. Rathi, P. Narayana Babu, S.A. Kuriakose, D.R.M. Samudraiah, & A.S. Kiran Kumar, "Methane sensor for Mars," *Current Science* 109, (2015), 1087–1096.

38 "US-India Joint Working Group on Civil Space Cooperation Joint Statement," *US Department of State - Archives*, Accessed on 14 July 2005, Bangalore, India, https://2001-2009.state.gov/p/sca/rls/pr/2005/49656.htm

39 "US, India to collaborate on Earth, Mars Missions," NASA Science – Mars Exploration Program, Accessed on 30 September 2014, https://mars.nasa.gov/news/1721/us-india-to-collaborate-on-earth-mars-missions/

40 "Volume 1: Mars Observer Mission Failure Investigation Board Report," *National Aeronautics and Space Administration*, 31 December 1993, http://athena.ecs.csus.edu/~grandajj/ME296J/4.%20Failure%20Reports/Mars_Observer_12_93_MIB.pdf

41 From National Aeronautics and Space Administration website, https://www.nasa.gov/centers/jpl/missions/mgs.html

42 From Mars Global Surveyor, Jet Propulsion Laboratory website, https://mars.nasa.gov/mgs/mission/spacecraft.html

43 A.L. Albee, R.E. Arvidson, F. Palluconi, & T. Thorpe, "Overview of the Mars Global Surveyor mission," *Journal of Geophysical Research* 106, (2001), 23291–23316.

44 J. Kawaguchi, "On the Lunar and Heliocentric Gravity Assist Experienced in the Planet-B ("Nozomi")," https://issfd.org/ISSFD_1999/pdf/IFL_3.pdf

45 "Mars Climate Orbiter Arrival – Press Kit," *National Aeronautics and Space Administration*, September 1999, https://www2.jpl.nasa.gov/files/misc/mcoarrivehq.pdf

46 "2001 Mars Odyssey Arrival – Press Kit," *National Aeronautics and Space Administration*, October 2001, https://mars.nasa.gov/odyssey/files/odyssey/odysseyarrival1.pdf

47 From European Space Agency website, https://sci.esa.int/web/mars-express/-/47364-fact-sheet

48 "Mars Reconnaissance Orbiter Arrival – Press Kit," *National Aeronautics and Space Administration*, March 2006, https://mars.nasa.gov/files/mro/mro-arrival.pdf

49 "India carves a unique place in space," *Press Information Bureau – Special Service and Features, Government of India*, 21 August 2017, https://pib.gov.in/newsite/printrelease.aspx?relid=170115

50 "Indian Space Research Organization – Annual Report 2016–2017," *Department of Space – Government of India*, https://www.isro.gov.in/sites/default/files/flipping_book/annualreport-eng-2017/files/assets/common/downloads/Annual%20Report%202016-17.pdf

51 E. Chassefière, O. Korablev, T. Imamura, K.H. Baines, C.F. Wilson, D.V. Titov, K.L. Aplin, T. Balint, J.E. Blamont, C.G. Cochrane, Cs. Ferencz, F. Ferri, M. Gerasimov, J.J. Leitner, J. Lopez-Moreno, B. Marty, M. Martynov, S.V. Pogrebenko, A. Rodin, J.A. Whiteway, L.V. Zasova, J. Michaud, R. Bertrand, J.-M. Charbonnier, D. Carbonne, P. Raizonville, & EVE Team, "European Venus Explorer (EVE): an in-situ mission to Venus," *Experimental Astronomy* 23, (2009), 741–760.

52 J.S. Greaves, A.M.S. Richards, W. Bains, P.B. Rimmer, H. Sagawa, D.L. Clements, S. Seager, J.J. Petkowski, C. Sousa-Silva, S. Ranjan, E. Drabek-Maunder, H.J. Fraser, A. Cartwright, I. Mueller-Wodarg, Z. Zhan, P. Friberg, I. Coulson, E. Lee, & J. Hoge, "Phosphine gas in the cloud decks of Venus," *Nature Astronomy*, (2020), https://doi.org/10.1038/s41550-020-1174-4

53 L.S. Glaze, C.F. Wilson, L.V. Zasova, M. Nakamura, & S. Limaye, "Future of Venus research and exploration," *Space Science Reviews* 214, (2018), 89.

54 "Indian Space Research Organization – Annual Report 2016–2017," *Department of Space – Government of India*, https://www.isro.gov.in/sites/default/files/flipping_book/annualreport-eng-2017/files/assets/common/downloads/Annual%20Report%202016-17.pdf

55 P. Bagla, "India eyes a return to Mars and a first run at Venus," *Science*, 17 February 2017, https://www.sciencemag.org/news/2017/02/india-eyes-return-mars-and-first-run-venus

56 L. Shekhtman, "NASA Goddard Team Selected to design concept for probe of mysterious venus atmosphere," *National Aeronautics and Space Administration – Goddard Space Flight Center*, 26 February 2020, https://www.nasa.gov/feature/goddard/2020/nasa-goddard-team-selected-to-design-concept-for-probe-of-mysterious-venus-atmosphere

57 N.A. Eismont, L.V. Zasova, A.V. Simonov, I.D. Kovalenko, D.A. Gorinov, A.S. Abbakumov, & S.A. Bober, "Venera-D mission scenario and trajectory," *Solar System Research* 53, (2019), 578–585.

58 *"ISRO to send first Indian into Spacce by 2022 as announced by PM, says Dr. Jitendra Singh. Rs. 10,000 crore mission will be a turning point in India's space journey; most engineering components are ready: ISRO Chairman. Chandrayaan-2 scheduled to be launched in January, 2019,"* Press Information Bureau – Department of Space, Government of India, 28 August 2018, https://pib.gov.in/PressReleseDetail.aspx?PRID=1544147

59 "Scientists Discuss Indian Manned Space Mission," *Indian Space Research Organization*, 7 November 2006, https://www.isro.gov.in/update/07-nov-2006/scientists-discuss-indian-manned-space-mission

60 "Space Capsule Successfully Recovered," *Indian Space Research Organization*, 22 January 2007, https://www.isro.gov.in/update/22-jan-2007/space-capsule-successfully-recovered

61 M. Natsuisaka, H. Ishizuka, T. Yamazaki, A. Higashibata, N. Ishioka, Y. Suwa, M. Ohmori, Y. Shimura, M. Nagase, M. Kamada, M. Seki, T. Hashizume, & S. Yoda, "International cooperation with India," *Journal of the Japan Society of Microgravity Application* 27, (2010), 158–164, https://www.jstage.jst.go.jp/article/jasma/27/3/27_158/_pdf

62 "Crew Module Atmospheric Re-entry Experiment (CARE)," *Vikram Sarabhai Space Centre, Indian Space Research Organization*, https://www.vssc.gov.in/VSSC/index.php/care

63 "Successful flight testing of crew escape system – technology demonstrator," *Indian Space Research Organization*, 5 July 2018, https://www.isro.gov.in/update/05-jul-2018/successful-flight-testing-of-crew-escape-system-technology-demonstrator

64 "French Agency CNES to Aid ISRO's Space Station Project," *IFCCI-CCI France-Inde – Communication - Press*, 27 January 2020, https://www.ifcci.org.in/news/n/news/french-agency-cnes-to-aid-isros-space-station-project.html

65 "Glavkosmos signed a contract for space training of Indian astronauts," *GLAVKOSMOS – News*, 1 July 2019, https://www.glavkosmos.com/en/glavkosmos-signed-a-contract-for-space-training-of-indian-astronauts/

66 "First meeting of Gaganyaan National Advisory Council," *Indian Space Research Organization*, 8 June 2018, https://www.isro.gov.in/update/08-jun-2019/first-meeting-of-gaganyaan-national-advisory-council

67 "Space in Parliament – Budget Session of Parliament 2019 (June – July 2019)," *Department of Space – Government of India*, https://www.isro.gov.in/sites/default/files/article-files/parliament-questions/budget_session_2019_-_english.pdf "The Global Exploration Roadmap," *International*

68 *Space Exploration Coordination Group*, January 2018, https://www.nasa.gov/sites/default/files/atoms/files/ger_2018_small_mobile.pdf

3 Dawn of the Second Space Age

The verity that the current space industry is an outcome of the geo-politics that resulted after the Second World War is valued rarely. Germany made great strides with satellite and spacecraft technologies, since the end of that War, but was bereft of space launch capabilities for many decades to only start it recently.[1] Australia[2] and New Zealand[3] accommodated the US' critical ground-based space assets but instituted their space agencies only recently. African nations that depended on Europe and the US for their Earth-observation and communication satellites now can avail the same services at cost-effective pricing from China and India. In the last 20 years, countries like Luxembourg, the United Arab Emirates (UAE), Israel, Singapore and Lithuania have birthed focused space programmes resourcefully. These births are signs of the coming Second Space Age.[4]

The First Space Age effectively ended during the second half of the 2000s. The end began when the US internet giant Google sponsored an international competition managed by the X-Prize Foundation known as the Google Lunar X-Prize. The competition was the first of its kind when a not-for-profit, with a private internet company's sponsorship, sponsored an international contest for private not-for-profit and for-profits to build, launch and operate a lunar lander and rover mission. It was perhaps the first time an entirely private venture was executing a cis-lunar mission without any direct connection with any space agencies.[5] This contest did not deliver a definite result, but a decade later, it kickstarted US-led international commercial interplanetary connectivity plans.

The ESA has ventured on a similar path. It is progressively commercialising robotic and human operations in low-earth orbit and on the Moon through an initiative known as Business in Space Growth Network (BSGN).[6] The BSGN is consorting with European start-up accelerators, government and national regulatory bodies of member states, venture capital firms and finance networks, business support

DOI: 10.4324/9781003152934-3

expertise, non-space industries and sectoral experts, and academic institutions. JAXA has initiated a commercialisation drive with some globally high-ranking electronics, automobile, precision instruments and finance companies. Called the JAXA Space Innovation through Partnership and Co-creation (J-SPARC),[7] this commercialisation drive gives the Japanese private sector a platform to innovate space technologies, applications and solutions. India's space reforms of May 2020 intend to take a commercialisation trajectory concomitant to its contemporaries. These shared undertakings point out again at the beginning of the Second Space Age. Regardless of this promising outlook, the path ahead for the world space economy is replete with ambiguity that must be forecasted meticulously.

New entrants in space exploration

Several countries that never possessed space ambitions during most years of the First Space Age began establishing space agencies since 2000. Unlike the established first-generation space agencies that undertook end-to-end innovation, design, operations and solutions provider, the newer crop act more like faciliatory regulators geared to assist their domestic private sector. The new-age space agencies cultivate space innovation ecosystems to eventually plug them into the global space economy and increase their nation's stakes. The newer crop comprehends that it cannot play an interventionist role like the space agencies of yesteryears. An interventionist role can arrest their nation's share in the global space economy and its speed of growth, ultimately affecting their comprehensive security matrix adversely.

Many third-generation spacefaring nations with advanced-economy status, such as Australia, New Zealand and the UAE, operate under the US' techno-security umbrella. Therefore, their space plans are congruent with the US' visions and goals about the interplanetary space. For example, the Australian government's 'Moon to Mars Initiative' programme objectives explicitly illustrates Australia's confluence with the US' interplanetary plans and aspirations.[8] Australia is investing $ 150 million to fuse itself in high-technology readiness level (TRL) space RDTE domains. These include robotics, automation, artificial intelligence, machine learning, optical communications, in-situ physicochemical investigations, space and remote medicine, next-generation digital remote sensing of planets, small Solar System bodies and satellites. Instead of working hastily on ostentatious technologies, Australia with the 'Moon to Mars Initiative' aims to use technology RDTE and commercialisation to enter into global space industry supply chains and dominate the new technology markets. This approach creates baseline space technology capability that gradually helps in national economic security. The 'Moon to Mars Initiative' is exceptional from the Australian government's other projects such as the 'International Space Investment Initiative' and the 'Space Infrastructure Fund.'[9] These two projects are part of the Australian Civil Space Strategy 2019–2028[10] that are focusing to develop suborbital and Earth-oriented remote sensing, situational awareness and communications capabilities interplanetary connectivity. With slight modifications, the objectives chalked by Australia also resonate with the long-term visions and goals of other new spacefaring nations.

The UAE's 'National Space Strategy 2030' aligns with its larger national ambitions as illustrated in the 'UAE Centennial Plan 2071.'[11] This strategic plan aims to make UAE a hub of globally leading space services, advance space R&D and manufacturing competence, nurture space innovation culture and ensure a supportive national legislative framework to fit the emerging global space economy's needs. These multifaceted goals fit well with the UAE's socio-economic target of creating the best education, economy, happiest society and best government in the world.

In 2018, the UAE signed an 'Implementing Agreement' with the US as they agree to cooperate in the domains of space exploration and human spaceflight in continuation of their close cultural, economic and diplomatic relations.[12] The UAE's Mars-bound *Hope* mission, launched in July 2020, has been accomplished with technical and scientific backing from US universities.[13] The UAE Astronaut Programme,[14] the UAE Mars Scientific City in Dubai,[15] the *Rashid* mission to the Moon scheduled for launch in 2024,[16] the Space on Earth Analog Mission to train the UAE astronauts[17] or the ambitious Mars 2117 megaproject[18] will have considerable inputs from the US and its global partners. The Joint Statement on the Launch of the US–UAE Strategic Dialogue,[19] released in October 2020, identified both civil and commercial space exploration and human spaceflight as an area of their bilateral partnership. The cooperation in these areas with the US can be viewed as the UAE's bid to ensure successful post-oil technology-driven diversification of its economy with a resilient and competent partner.

Although the US might be the only driver of interplanetary infrastructure build-up, many new entrants are working with other power centres. For example, Luxembourg has become an epicentre of space exploration, space resources extraction and space resource utilisation within the European Union within a short period. The Grand Duchy has attracted numerous start-ups in these domains, including satellite construction, space-based robotics and space-based artificial intelligence originating from countries worldwide, including non-European countries like Japan to the US. This new 'space resources' ecosystem was possible due to the Grand Duchy's pre-emptive adoption of a legislation titled the 2017 Act on the Exploration and Use of Space Resources, better known as the Space Resource Act.[20] This legislation is one of the first legal and regulatory frameworks adopted by any country for authorising and supervising protocols and procedures for space missions aiming to explore, extract and utilise natural resources from planets, natural satellites and small Solar System body surfaces.

Since space resources extraction and utilisation, if not handled carefully, can potentially vortex into a politico-legal spiral, many countries even today have avoided it. Luxembourg went one step ahead and is currently in a funding agreement with the United Nation's Office of Outer Space Affairs (UNOOSA) to support the latter's 'Space Law for New Space Actors' project.[21] With this project, the UN member states can receive tailored assistance in drafting national space policies and legislation in tune with global space treaties and laws, thereby fostering sustainability of economically driven space activities. Furthermore, Luxembourg's proactive role in the space resource utilisation domain has spawned a new ESA research centre, known as the European Space Resource Innovation Centre (ESRIC), in November 2020. The ESRIC is now an ESA nodal body dedicated to the next-generation European robotic missions to asteroids, Mars, and the Moon, including its human-habitation megaproject on the lunar surface, Moon Village.[22] While continuing to be an important member of the ESA, Luxembourg has also agreed to be part of the US' Artemis Programme for sustained human presence on the Moon.[23]

These are spirited examples of some emergent space players. However, not all emergent space players possess similar grand plans nor have the bandwidth to execute them. Many new entrants will want to advance their space capabilities slowly and steadily and under the sway of their geopolitical umbrella. For instance, the Egyptian space programme has been particularly close to Russia. The Nigerian space programme has benefited from China's support. Despite having neighbours such as India and China with solid space competence, Bangladesh prefers a US partnership. These preferences signify the long-term alliances these countries want to pursue with the current and touted superpowers.

Industry 4.0 galvanising the end of the First Space Age

The Fourth Industrial Age is the product of a globalised world. From 3D printing to artificial intelligence, from pattern recognition to robotics, from optical communications to electric propulsion, the world is now brimming with immense possibilities of Industry 4.0 technologies. Unlike the earlier-era Industry 3.0 technologies, which were researched and developed in academia and large technology companies, most Industry 4.0 technologies come out of start-up ecosystems, small- and medium-scale enterprises and flexible non-governmental financing mechanisms. During the 20th century, high-end technologies were developed within national borders' confines and funded only by the governments. In the Fourth Industrial Age, cutting-edge technologies come to the fore through transnational venture capital and a promise of an international market for them. Consequently, private monetary wealth in the form of equity-based investments, fund-of-funds, angel investments, venture capital and initial public offerings are spawning Industry 4.0 technologies (refer to Figure 3.1).

As witnessed by the examples of Luxembourg, the UAE and Australia, national governments are pushing scientific frontiers through public–private synergies and acting as enablers and not nannies. This change of attitude is also being witnessed in the space sector, which has primarily been a government-protected domain. The end of road for First Space Age technologies began creating a vacuum in the US space programme, much before any other country. The US began sensing stagnancy in their space programme with the few and entrenched space contractors. To avert this stagnancy, NASA first came up with the Commercial Orbital Transportation Services (COTS) programme in 2006,[24] the Commercial Resupply Services (CRS) in 2008[25] and the Commercial Crew Development (CCDeV) programme in 2011.[26] These programmes brought new, innovative and agile space contractors to the fore. Today, many new space contractors win contracts for space launch services, logistics supplies and commercial human spaceflight to the International Space Station. The new contractors brought in new and lean manufacturing practices, made space contracts economically feasible and generated market possibilities to shut out aging space technologies and replace them with future-looking ones.

The strong brand-perception of new companies like SpaceX and Blue Origin and the organisational reforms undertaken by older space contractors like Boeing, Energia, Airbus Defence and Space have been

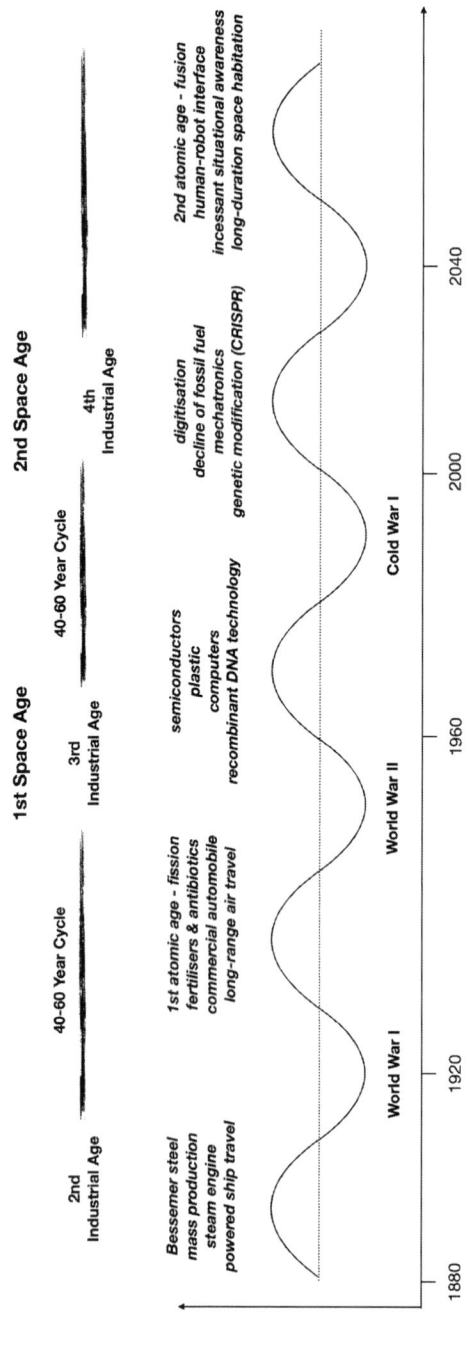

Figure 3.1 The crest and trough of wars and technology revolutions and the concomitance of Fourth Industrial Age and Second Space Age

well regarded. This acknowledgement further pushed the US government to pass the Commercial Space Launch Competitive Act of 2015,[27] which pledges to support private aerospace competitiveness and entrepreneurship, particularly in the pursuit of commercial exploration outer space and exploitation of inanimate natural resources from space. Although the US seems committed to maintaining its supremacy in outer space technologies, it has also wisely chosen a band of new spacefaring countries that will partner with it in the same pursuit. These countries are sheltering specific Industry 4.0 technologies that they can champion across various industries. Through them, they can equitably contribute to the US-led commercial space exploration ventures under the Artemis Accords initiated in 2020.

With a greater emphasis on Industry 4.0, space programmes worldwide are inevitably witnessing a sea change in terms of technologies deployed. This transformation widened the horizon of space-based activities constricted to few spacefaring countries during the First Space Age.

Consequential technologies of the Second Space Age

The US government's pledge to encourage and promote commercial space activities has been well taken up by numerous countries worldwide. Some of these countries align with the US and are part of the Artemis Accord; some, like India, Germany and France, are US' non-aligned space partners but not a party to the Artemis agreement. China has taken inspiration but has modified it according to its political characteristics. The comprehensive global support for commercial space activities is creating an enormous potential for the global space economy. Nations approving of this new potential are facilitating space startups, medium-scale enterprises, large private conglomerates and state-led companies. The commercialisation is not only making the vendor and supplier directory voluminous but also expanding the space innovation and industry ecosystems. It gives equal opportunities for companies with less or no experience in the First Space Age space industry.

Offering an equal pedestal to private entities across the spectrum is an efficient way to tap into the diverse innovations emerging. For instance, space start-ups often have new, bold and radical proofs of concept and prototypes. Since they usually grow out of academic environments, their product offerings are at low-technology readiness levels, yet they can offer high return on investment. Medium-scale space innovation enterprises are early implementers of newly innovated and proven subsystems, and in the recent past, have also commercialised scaled-down versions of existing technologies. For example, numerous small launch vehicle, small satellite and subsystem companies belong to the medium-scale innovation enterprise cohort. Large private and state-owned space contractors have greater market heft to carry out high-risk-high-reward space projects. However, they are contingent on national authorisation as most of such projects are part of the government's manifesto and long-term bipartisan national strategy.

In the pre-commercialisation era, many countries would assign their space agencies to carry out innovation through public funds and then transfer technology to a select group of vendors for contract manufacturing. That model has gone past its expiry. Today, space agencies' role is left mainly as a regulator, facilitator, and doers of non-commercial but strategic space projects. With a reinforced regulatory appointment, a space agency now can assign businesses to all the innovating, manufacturing and service-providing entities, monitor their output, and maintain technology standards. Therefore, the technologies consequential in the Second Space Age will come from a broad innovation base than it ever came earlier. This broad base will have a multiplier effect on the global space economy. It will involve many more nations and the vast and multiplying innovation and industrial ecosystems within them.

The private space ecosystems will increasingly find themselves demonstrating proofs of concept of interplanetary exploration and connectivity technologies funded through financial grants from space agencies, venture funding and trans-industrial fund of funds. Once these proof of concepts graduate to higher-technology-readiness levels, they will find mercantile space applications. With interplanetary connectivity becoming an economic engine of deep-space activities, many of these ecosystems have begun to seek business sustenance by getting contracts for commercial operations in the Earth's orbits, including orbital contractual RDTE, orbital manufacturing services and tourism. They anticipate contracts for laying interplanetary telecommunications infrastructure and operations, particularly between Earth, Moon, Mars and Lagrange Points (Table 3.1).

There is a growing interest in companies that could develop reusable, cost-effective and safe interplanetary transportation infrastructure for logistics supply, construction, robotic deployment and crewed habitats on Moon and Mars's surface. Many of the technologies coming out of these space ecosystems are designed to have tremendous commercial applications across various industries and non-space

Table 3.1 Short-term objectives of interplanetary connectivity infrastructure

Short-term Objectives from Interplanetary Infrastructure	Global Space Supply Chain Integration	Novel Space Technology Demonstration	Space Technology Leadership
International Objectives			
Expand the number of domestic companies in global supply chains and future space markets	✓		
Demonstrate national capabilities on global scale	✓	✓	✓
Sustainably integrate into global supply chains	✓		
Develop nationally conducive international code for responsible conduct in outer space	✓	✓	✓
Domestic Objectives			
Accelerate space economy targets and generate high-skill employment	✓	✓	✓
Develop flagship space capabilities strengthening national stakes in international space exploration industry		✓	✓
Ensure participation of wide range of spin-in industries in space economy	✓	✓	✓
Consistently maintain supply of employable and skilled human resource	✓	✓	✓

applications. The expansion of the market triggered in these early years of the Second Space Age will portend all those entities associated with commercial space activities that have lost the ability to adapt. The expansion of the ecosystems associated with space could result in companies contesting vigorously for business. Since the space economy's magnitude is enormous, the competition will be healthy with an immense business greenfield always available (Table 3.2).

Table 3.2 Applications of various technologies in the interplanetary connectivity infrastructure

Technology Products	Interplanetary Connectivity Applications
Remote sensing payloads	Surveillance, reconnaissance and prospection of natural resources on extraterrestrial surfaces
Surface telecommunication equipment	Internet data transmission, internet of things command and control, remote navigation
Surface transportation systems	Robotic exploration rovers, human-rated roving vehicles, logistic and human-rated landers
Interplanetary transport and traffic management systems	Interplanetary logistics and human-rated transportation vehicles, deep space communication relay systems
In-situ resource utilisation and extraction and clean-fuel depots	Vehicular refuelling; power, heating and life-support utilities
Space-based electronics and semiconductors	Radiation-hardened electronics and semiconductors for all space and extraterrestrial applications
Unconventional energy and propulsion systems	Hybrid-fuel, ion propulsion, nuclear-plasma/ solar sails propulsion for launch vehicles, satellite and spacecraft propulsion, respectively.

The rise and rise of private sector in space exploration

The global space exploration pursuits are steadily moving from being agency-driven to now being industry-driven. The credit for this transition goes to the advanced spacefaring democracies of the world. The space agencies of non-socialist countries long ago understood the impact of sourcing technologies innovated by the private space sector despite inventing world-class technologies in their government-run laboratories.

In recent years, China, being a Communist state with absolute control over all business entities, has permitted numerous quasi-private spaces, aerospace and defence companies. These quasi-private companies supplement the backbone of China's high-technology space industrial ecosystem that the Communist Party of China (CPC) controls. The party controls the ecosystem via the State-owned Assets Supervision and Administration Commission of the State Council (SASAC), the Ministry of Industry and Information Technology

(MIIT) and the Central Military Commission of the People's Republic of China. The SASAC is the largest economic entity in the world, with cumulative assets surpassing 26 trillion US dollars. The MIIT oversees the China National Space Administration (CNSA) – China's civilian space agency, the Central Military Commission of the People's Republic of China and the Central Military Commission of the Communist Party of China.

China has made tremendous forays with its space programme using capacities existing entirely within its state apparatus mentioned above. The SASAC companies, the CASTC and the China Aerospace Science and Industry Corporation (CASIC) have been mainly responsible for end-to-end manufacturing services for the Chinese space programme. Together they have built the Chinese launch vehicles, spacecraft, space capsules, landers, rovers, satellites and the Chinese space station modules. Despite a strong product portfolio, the Chinese state feels the need to have a band of quasi-private space, aerospace and defence companies superintended by the CASTC, the CASIC and other SASAC-controlled state companies.

The Chinese have many reasons to keep such quasi-private companies. The foremost is tapping global talent. China is aiming to become a powerful magnet for a highly skilled workforce from around the world. It understands that howsoever massive, the SASAC will be unable to attract foreign talent due to its close interfaces with the Communist Party of China. The divide between the Chinese civilian and military sectors is indistinct, and the SASAC companies, therefore, are inundated with sensitive projects and intelligence. The Chinese tact to attract talent has created quasi-private ecosystems of high-technology companies in Beijing, Wuhan, Shanghai, Harbin and Shenzhen. These ecosystems mimic many of the Western innovation ecosystems and additionally offer attractive remunerations. These factors attract overseas non-national talent and create a professional environment similar to the West to address the aspirations of its native talent. However, deep within the corporate structure are the signatures of the government control over these companies.

For instance, ExPace Technology Corporation, a supposed private space-launch company started in 2016, is a subsidiary of the CASIC.[28] Likewise, many other Chinese space-launch companies like LinkSpace, LandSpace, and Beijing Interstellar Glory are university-run start-ups. They have substantial governmental stakes in them, mainly of the SASAC companies. This phenomenon is repeated in the satellite and spacecraft business verticals too. Chinese quasi-private space companies are minuscule in market valuations, possession of ground infrastructure and product portfolio expanse compared with the SASAC companies. In this way, the State Council seeks to create small and semi-autonomous spirited centres within the extensive apparatus of the SASAC companies that could repurpose existing SASAC technologies and commercialise them.

Russia has been consistently supporting quasi-private and semi-autonomous private companies emerging from its Soviet-era 'design bureaus.' In the post-Soviet period, Russia converted many of these design bureaus into state-controlled corporations, some as public joint-stock corporations and some as closed joint-stock corporations. Since 2009, Russia has nurtured a few space start-ups through the private Skolkovo Innovation Fund[29] located in the Skolkovo Innovation Center near Moscow. Russia's slow but sure-footed privatisation initiatives have shown tremendous promise since then. Private Russian companies like *Dauria* and *Sputnix* today operate microsatellites, subsystem company, like *RoboCV*, build computer vision for lunar rovers and warehouse robots, *Apis Cor Engineering*, designs a 3D-printed Martian habitat. However, its approach towards privatisation is less ambitious than China, owing to its comparatively weaker economy. The emphasis given by Russia and China, two Communist states, to the commercialisation of the space sector is a telltale of non-governmental innovation and manufacturing's growing significance.

However, the credit of the private sector's growing emphasis goes to the space policy decisions taken by private innovation friendly space agencies at regular intervals. The democratic world was the first to employ the creative talent residing in the private sector. Some of the world's most successful private space innovation and manufacturing companies reside in these democratic countries. These companies are a model for communist countries and many other democracies and monarchies vying to make their mark with niche capabilities and acquire a commanding stake in the global space economy.

Notes

1 M. Fuchs, A spaceport in the North Sea would catapult Germany into the future as a location of the aerospace industry, *OHB Systems*, 2 October, 2020, https://www.ohb.de/en/magazine/space-encounter-a-spaceport-in-the-north-sea-would-catapult-germany-into-the-future-as-a-location-of-the-aerospace-industry

2 "New satellite ground station in WA fully operational," *Department of Defence – Australian Government*, 12 October 2020, https://www.minister.defence.gov.au/minister/lreynolds/media-releases/new-satellite-ground-station-wa-fully-operational

3 M. Blenkin, "US military provides satellite communications support for ADF in bushfire missions, *Space Connect*, 29 January 2020, https://www.spaceconnectonline.com.au/operations/4106-us-military-provides-satellite-communications-support-for-adf-in-bushfires-mission

4 C. Giri, "A Space Exploration Industry Agenda for India," *Gateway House – Indian Council for Global Relations*, May 2020, https://www.gatewayhouse.in/wp-content/uploads/2020/05/26-May_A-space-exploration-industry-agenda-for-India_Chaitanya-Giri_Final.pdf

5 V. Shammas, T.B. Holen, "One giant leap for capitalistkind: Private enterprise in outer space," *Palgrave Communications* 5, (2019), 10, https://www.nature.com/articles/s41599-019-0218-9?fbclid=IwAR0txc-kU2U2guKLG ci84YNJNV_2RJy2wXPKNEiu2BR6noK25hRa5Zhf8Lc

6 "The Business in Space Growth Network," *European Space Agency*, https://www.cosmos.esa.int/documents/4100099/4100114/ BSGN+overview.pdf/5ef281de-f74a-22d4-0266-b6d3acdb4c02?t= 1589789310007

7 "The Business in Space Growth Network," *European Space Agency*, https://www.cosmos.esa.int/documents/4100099/4100114/ BSGN+overview.pdf/5ef281de-f74a-22d4-0266-b6d3acdb4c02?t= 1589789310007

8 Retrieved from the J-SPARC website, https://aerospacebiz.jaxa.jp/solution/ j-sparc/

9 "Moon to Mars Initiative: Launching Australian industry to space," *Department of Industry, Science, Energy and Resources – Australian Government*, 16 February 2021, https://www.industry.gov.au/news/ moon-to-mars-initiative-launching-australian-industry-to-space

10 "Scrutiny of Commonwealth expenditure," *Department of Industry, Science, Energy and Resources – Parliament of Australia*, 27 May 2020, https://www.aph.gov.au/Parliamentary_Business/Committees/Senate/ Scrutiny_of_Delegated_Legislation/Scrutiny_of_Commonwealth_expenditure/Portfolios/Industry

11 H. Buamim, "Industry Perspective: The sky is no longer the limit," *Global Governance Project*, 13 November 2020, https://www.globalgovernanceproject.org/industry-perspective-the-sky-is-no-longer-the-limit/

12 G. Hautaluoma, "NASA, UAE Space Agency Sign Historic Implementing Arrangement for Cooperation in Human Spaceflight," *National Aeronautics and Space Administration*, 4 October 2018, https://www.nasa.gov/press-release/nasa-uae-space-agency-sign-historic-implementing-arrangement-for-cooperation-in-human

13 "UAE's Hope mission on its way to Mars," *Ann and H.J. Smead Aerospace Engineering Sciences – College of Engineering and Applied Science – University of Colorado Boulder*, 20 July 2020, https://www.colorado.edu/ aerospace/2020/07/20/uaes-hope-mission-its-way-mars

14 Retrieved from the Mohammed Bin Rashid Space Centre, https://www.mbrsc.ae/astronaut-programme

15 "Policy in Action – Mission to Mars," *Federal Competitiveness and Statistics Authority – Government of the United Arab Emirates*, (2019), https://fcsa.gov.ae/en-us/Lists/D_Reports/Attachments/26/Issue%20 11%20-%20Mission%20to%20Mars.pdf

16 E. Gibney, "UAE ramps up space ambitions with Arab world's first Moon mission," *Nature*, 5 November 2020, https://www.nature.com/articles/d41586-020-03054-1

17 B. Alfeeli, "#SpacewatchGL Opinion: My 2020 in a Review – By Bassam Alfeeli," *Spacewatch Global*, https://spacewatch.global/2020/12/review-bassam/

18 "UAE Future – 2030–2117," *The United Arab Emirates Government Portal*, https://u.ae/en/more/uae-future/2030-2117

19 "Joint Statement on the Launch of the US-UAE Strategic Dialogue," *Media Note – Office of the Spokesperson, US Department of State*, 22 October 2020, https://2017-2021.state.gov/joint-statement-on-the-launch-of-the-u-s-uae-strategic-dialogue/index.html

20 "International Space Law – National Space Law," Luxembourg Space Agency, https://space-agency.public.lu/en/agency/legal-framework.html

21 "United Nations Office for Outer Space Affairs signed an agreement with the Government of Luxembourg to launch new "Space Law for New Space Actors" project," *Press Release – United Nations Office for Outer Space Affairs*, 13 November 2019, https://www.unoosa.org/oosa/en/informationfor/media/2019-unis-os-523.html

22 "Luxembourg teams up with ESA to create a unique "European Space Resources Innovation Centre" to be established in the Grand Duchy," Press Release – The Luxembourg Government, 18 November 2020, https://gouvernement.lu/en/actualites/toutes_actualites/communiques/2020/11-novembre/18-luxembourg-spaceresources.html

23 "Luxembourg, NASA and several other partner countries are among the first signatories of the Artemis Accords," Press Release – The Luxembourg Government, 14 October 2020, https://gouvernement.lu/en/actualites/toutes_actualites/communiques/2020/10-octobre/14-luxembourg-nasa-artemis.html

24 "Commercial Orbital Transportation Services – A New Era in Spaceflight," National Aeronautics and Space Administration, May 2014, https://www.nasa.gov/sites/default/files/files/SP-2014-617.pdf

25 "NASA Awards Space Station Commercial Resupply Services Contracts," National Aeronautics and Space Administration – Contract Release: C08-069, 23 December 2008, https://www.nasa.gov/home/hqnews/2008/dec/HQ_C08-069_ISS_Resupply.html

26 "NASA Awards Next Set of Commercial Crew Development Agreements," National Aeronautics and Space Administration – Contract Release: 11-102, 18 April 2011, https://www.nasa.gov/home/hqnews/2011/apr/HQ_11-102_CCDev2.html

27 "H.R. 2262 – US Commercial Space Launch Competitiveness Act," Public Law No.: 114-90, 114th Congress Public Law 90, US Government Publishing Office, https://www.congress.gov/bill/114th-congress/house-bill/2262/text

28 P. Keane, "ExPace, China's very own SpaceEx," *Asian Scientist*, 20 September 2016, https://www.asianscientist.com/2016/09/columns/final-frontiers-expace-chinas-version-spacex-casic/

29 A. Ilyin, "Russian business: a long road to the stars for private space initiative," ROOM Journal, (2015), Issue #3(5), https://room.eu.com/article/Russian_business_a_long_road_to_the_stars_for_private_space_initiative

4 The global planetary exploration roadmap

Cooperation and contest

The world's foremost spacefaring superpowers have typical traits. They form strong space synergies with their geopolitical allies. The end goal of these space synergies is to graduate geopolitical alliances into astropolitical blocs. Any space industry entity residing outside this bloc's ambit is seen as a competitor or even a threat. Such astropolitical blocs seldom compete to cooperate than cooperate to compete when it comes to the RDTE and commercialisation of strategic space technologies.

For instance, the signatories of the US-led Artemis Accords cooperate with each other to compete with the small Chinese and Russian blocs. The Russian and Chinese blocs compete with each other and seldom cooperate. Likewise, there are neutral countries like India and France, which are not part of any astropolitical blocs due to their strategic autonomy. However, they do compete in an underplaying manner.

Unlike the US and China that maintain scant space diplomatic relations, the US and Russian space diplomatic relations have cooperated during their geopolitical contest. The US and Russia have deep and multi-decadal collaborations in their human spaceflight projects and about heavy-lift launch vehicles. The Artemis Accords are comparable to the *Interkosmos* programme of the Soviet Union. *Interkosmos* had COMECON countries and outlier partners like France and India. NASA's China Exclusion policy prevents any collaborative handshake with the Chinese space agency, the CNSA.[1]

The Communist Party of China (CPC) and Central Military Commission's direct linkages with the CNSA prevent any civilian space cooperation between the US and China. There are numerous platforms where major space agencies coordinate their planetary exploration

DOI: 10.4324/9781003152934-4

strategies and use them for their space diplomacy commitments. These include the UNOOSA, the Committee on Space Research (COSPAR), the International Astronautical Federation (IAF) and the recently announced Space20 grouping under the Group of Twenty (G20) multilateral. UNOOSA monitors global consensus for the United Nations Committee on the Peaceful Uses of Outer Space through treaties like the Outer Space Treaty of 1967. The other multilateral bodies primarily focus on space science and engineering domains. However, the one body that warrants detailed analysis is the International Space Exploration Coordination Group (ISECG).

Catalytic role of International Space Exploration Coordination Group

The ISECG was established in 2006 when 14 national space agencies decided to co-develop a voluntary and non-binding mechanism. Each of them exchange information regarding their interests, aims and preparations for space exploration. The ISECG is a pacific grouping for member agencies to share their roadmaps, discoveries, recommendations and outputs. Their space exploration strategies are concomitant with each other. With more countries intensifying their space exploration activities, the ISECG has grown to accommodate 26 space agencies from across the world.[2]

The main focus areas of the ISECG are long-term human and robotic exploration and presence on Mars, Moon, in their orbits, and in the cislunar (Earth-Moon) and interplanetary (Earth-Mars) space. However, for all practical matters, the ISECG is also becoming an instrument of multilateral space diplomacy and is stimulating space economy engagement. The ISECG has further elaborated joint engineering and scientific targets for all the member space agencies to attain these objectives. With these focus areas, the ISECG intends to lay a common framework for the member space agencies to cooperate, coordinate and collaborate.

The ISECG regularly updates a Global Exploration Roadmap (GER)[3] that maintains a roster of the most critical technologies needed for planetary exploration (Table 4.1). These technologies are not off-the-shelf products but will need constant R&D and conveyor-belt manufacturing once Mars- and Moon-bound missions become frequent. Therefore, this long list has become a primer for the private space industry.

Table 4.1 Technologies identified by the ISECG Global Exploration Roadmap for deployment in the Inner Solar System

Important GER Technologies	Lunar Vicinity	Lunar Surface	Mars Vicinity	Mars Surface
Strong Thermal Protection System and Heat Shield (for multiple atmospheric entry in a single mission)			✓✓	✓✓
Thermal Control		✓		✓
Orbital and Surface Cryogenic Propellant and Fuel Storage		✓	✓	✓
Low-temperature mechanisms	✓	✓	✓✓	
Inflatable module structure and materials			✓✓	✓✓
Entry, Descent and Landing modules for heavy payloads on Mars				✓
Regolith and dust mitigation		✓		✓
Mission control automation beyond low-Earth orbit			✓	✓
Surface transportation systems		✓		✓
Rapid-access extra-vehicular activity		✓✓	✓✓	✓✓
In-situ Resource Utilisation on Mars				✓
Galactic Cosmic Rays radiation protection			✓	✓
Solar Particle Events radiation protection	✓		✓	✓
fire detection, prevention and low-pressure suppression	✓	✓	✓	✓
In-flight life-support system	✓	✓	✓	✓
Deep space human factors, ergonomics and habitability			✓	✓
Microgravity biomedical countermeasures, behavioural health and performance and medical care			✓	✓
Mars surface suiting				✓
Lunar surface suiting		✓		

(*Continued*)

Table 4.1 (*Continued*)

Important GER Technologies	Lunar Vicinity	Lunar Surface	Mars Vicinity	Mars Surface
Deep space suiting	✓		✓	
Closed loop life-support system			✓	✓
High data-rate low-latency downlink (hybrid RF-optical communications)			✓	✓
In-space navigation and timing for autonomous operations			✓	✓
High data-rate, inter-networked adaptive proximate telecommunications			✓	✓
High data-rate in-flight uplink communications			✓	✓
Beyond low-Earth orbit crew autonomy	✓	✓	✓	✓
Autonomous rendezvous and docking	✓		✓	
Autonomous vehicle systems management	✓	✓	✓	✓
Robots operating along with crew in spacesuits	✓	✓	✓	✓
Telerobotic control	✓	✓	✓	✓
Precision landing with obstacle and hazard avoidance		✓		✓
Low-temperature operating long-duration energy storage systems		✓		
High-specific surface transportation battery packs		✓		✓
Nuclear power for electric propulsion			✓ ✓	
Fission power for surface missions				✓ ✓
Autonomous high power in-space arrays			✓ ✓	
High-strength solar arrays for surface power supplies	✓ ✓			✓ ✓

(*Continued*)

Table 4.1 (*Continued*)

Important GER Technologies	Lunar Vicinity	Lunar Surface	Mars Vicinity	Mars Surface
In-space cryogenic liquid acquisition	✓ ✓			✓
Nuclear thermal propulsion engine			✓ ✓	
Electric propulsion and power processing	✓ ✓		✓ ✓	
Liquid oxygen/methane cryogenic propulsion		✓	✓ ✓	✓

Notes: ✓ – low-readiness levels of technologies; ✓ ✓ – high readiness levels of technologies.

The GER has been influential in building confidence in member space agencies to employ the private sector to develop crucial space technologies. The GER has enthused the Japanese private automaker Toyota to build a lunar surface transportation system, deemed LUNAR CRUISER, co-developed with the Japanese space agency, JAXA[4]. NASA is currently collaborating with the Battelle Energy Alliance (a partnership between AECOM, Electric Power Research Institute and BWXT) towards demonstrating nuclear fission reactor on the lunar surface by 2027.[5] The Finnish telecommunications giant Nokia is aiming to set up a fourth-generation long-term evolution (4G-LTE) infrastructure on the surface of the Moon.[6]

The ISECG is no détente platform for opposing geopolitical blocs. It is evident that NASA and the CNSA, despite their ISECG membership, have not used it to thaw their relations. NASA has not pulled back its China exclusion policy. Neither has the single-party-driven China been able to instill confidence. Similarly, ISRO and the CNSA have also continued to be distant from each other. However, with more nations pushing for their space agencies to join the ISECG, this grouping will begin to witness geopolitical-multipolarity in outer space.

Breeding of various astropolitical blocs

The ISECG's GER is an essential primer of the joint technological goals of member space agencies. However, the commonality of the ISECG identified goals does not attribute to the space agencies' common interests and the governments they represent. The ISECG is certainly a non-binding body where space agencies collaborate on the convergences in their lunar and Martian exploration endeavours.

However, the space agencies are committed to serving their divergent strategic national interests.

Russia pursues a two-prong strategy for planetary exploration as stated in its '*Principles of State Policies of the Russian Federation in the field of space activities up to 2030 and beyond.*'[7] Russia's two-prong strategy ensures its space institutions engage in space diplomacy yet contribute to building cis-lunar infrastructure through indigenous efforts. Roscosmos maintains a strong partnership with the ESA, German Aerospace Center (DLR), the CNES and NASA in RDTE of long-duration human spaceflight at the Scientific International Research in Unique Terrestrial Station (SIRIUS) at the Institute of Biomedical Problems.[8] Roscosmos and the various Russian space contractors (design bureau) have been the mainstay of crew flights and logistics to the International Space Station. However, this strong patronage has not resulted in Russia participating in the US-led international Artemis Accords.

The US initiated the Artemis Accords in October 2020 with ten national space agencies as to its signatories. The agreement is in principle based on the Outer Space Treaty of 1967. The Artemis Accords create a *de facto* alliance of space agencies to realise and upheld common strategic interests in outer space. The signatories are Australia, Brazil, Canada, Japan, Luxembourg, Italy, Ukraine, the United Kingdom South Korea, New Zealand and the UAE. The Artemis Programme is an international multicultural human spaceflight programme bound for the Moon but only limited to the Artemis Accords alliance members.

Although the Artemis Accords are new on the table, there is a conspicuous absence of Germany, France and India. These three countries have considerable heft in planetary exploration. Their absence alludes to the presence of the fourth category of intrepid space agencies that are married to their national ambitions than those of any astropolitical bloc. These three are not necessarily antagonistic to the US, Russian or Chinese plans. Nevertheless, they exude the confidence to collaborate based on their national interests. As mentioned earlier, nations do not compete to cooperate, but they do cooperate to compete. Each country possesses a certain degree of ambitions from space exploration. A nation's economy's heft, geopolitical clout, technological competence, supply of skilled human resources and innovation capital determine its ambition.

Where Russia, France, Germany and India appear pragmatic with their space ambitions, the US and China have engaged in an intense contest to dominate the low-Earth orbit, cislunar space and the lunar surface by the year 2040. China has made strides in the past two decades. The *Shenzhou* crewed spaceflight programme's success has

Table 4.2 Objectives of the space missions of the Chinese Lunar Exploration Program

Technical Objective	CLEP Missions
Orbiting the Moon	Chang'e 1 (2007) and Chang'e 2 (2010)
Landing on the nearside of the Moon	Chang'e 3 (2013)
Landing on the far-side of the Moon	Chang'e 4 (2019)
Sample-return missions from the Moon	Chang'e 5 (2020) and Change' 6 (unscheduled)
Establishing a robotic lunar surface research station	unscheduled

resulted in China registering its presence in the low-Earth orbit with the *Tiangong* technology-demonstrator space station in the late 2010s. China is preparing to construct a full-fledged and operational 'Chinese Large Modular Space Station' starting 2021 and is aiming to send Chinese crew on the Moon by 2036.[9] Since its inception in 2007, the Chinese Lunar Exploration Program (CLEP) has achieved four of the five technological objectives that it had envisioned when the programme began (Table 4.2).

Although China is not formally leading an alliance like the US' Artemis Accords, their crewed spaceflight and space station programme has roused interest in countries like Pakistan and Iran.[10] The China Manned Space Agency (CMSA) is also utilising its heft in the United Nations Security Council and the UNOOSA to reach out to UN member states and invite them to participate in its space station programme.[11] With this capacity-building exercise, China aims to nucleate soft-power gains from countries, particularly from the developing world, and perhaps develop an alliance equivalent to the Artemis Accords but with CPC's characteristics.

There are four astropolitical blocs in the world today. The US has recently formalised one bloc with the Artemis Accords, which contains some of its Five Eyes partners and its strategic military and economic partners. Beijing is looking to lead its bloc with geopolitical rivals of the Artemis Accords and those deeply integrated within the Belt and Road Initiative (BRI). European countries operate as an ESA-led bloc but with each ESA member possessing the autonomy to engage their national space agencies with counterparts outside the ESA. Russia, too has its band of reliant partners like Belarus, Egypt and Uganda, along with some yesteryears Eastern Bloc members. Interestingly, India is no member of any astropolitical bloc neither has it shown signs of raising its bloc. However, India fluidly engages in limited partnerships with various bloc leaders and members.

For instance, ISRO has limited relations with the CNSA. With the cold India–China relations, a space collaboration is nowhere on the anvil. However, with the CMSA-UNOOSA mechanism, the Indian Institute of Astrophysics (IIA) and the Indian Institute of Technology Banaras Hindu University (IIT-BHU) have been selected as experimenters on the China's Large Modular Space Station. They partner with the Russian Academy of Sciences and the Université Libre de Bruxelles, Belgium.[12] These collaborations are unconnected with the Indian Human Space Flight Programme (IHSFP) or the Indian space station mission scheduled for 2030.

Rationale of astropolitical bloc leaders

The astropolitical blocs have goals that surpass pageantry and propaganda. Each bloc is ideologically aligned and conjoined by common economic and security interests. An astropolitical bloc shares financial mechanisms for space projects, exchanges skilled human resources and maintains intergovernmental regulatory oversight. It also enters into technology-sharing agreements, upkeep space technology standards within the bloc, and propagate them outside the bloc. Bloc members cooperate within the coalition to compete with other astropolitical blocs.

Space exploration, therefore, is not entirely a scientific pursuit and is increasingly becoming an astropolitical connectivity pursuit for increasing the stakes of the bloc members. The astropolitical blocs share vendors, contractors and turnkey project managers. They assure sustenance or continuity of business to all these entities, thereby maintaining their edge as some of the world's most innovative and high-technology manufacturing companies. The interplanetary connectivity plans of the astropolitical blocs are not limited to the conventional bread-and-butter 'satellite and rocket' industries. For example, the ESA has roped in the US-based urban planning company Skidmore, Owings & Merrill and the Massachusetts Institute of Technology to design its lunar habitat known as the Moon Village.[13] Likewise, the China Aerospace Science and Technology Corporation (CASTC) is working on the Chinese government's proposed investment of 10 trillion dollars. The plan is to construct an Earth-Moon Special Economic Zone by 2050 and set up a lunar habitat close to the lunar south pole Aitken Basin.[14] The UAE's plans for setting up a Mars City 2117, with substantial US support, aim to establish building blocks of interplanetary connectivity.

Although interplanetary connectivity projects appear outlandish and impractical, the stock markets do not think so. For instance, companies like Lockheed Martin, Boeing, Airbus, Virgin Galactic are public listed in stock markets. However, of late, there have been a few space-industry-specific exchange-traded funds (ETFs) that have successfully made it in the stock markets. One such is the Standard and Poor's Depository Receipt (SPDR) S&P Kensho Final Frontiers ETF initiated in October 2018.[15] This Kensho Final Frontiers ETF allocates funds into aerospace and defence, semiconductors, information technology, electronic manufacturing, industrial machinery, construction and engineering, electronic components and metal and glass container entities. The Procure Space ETF, initiated in April 2019, has companies like Loral, Virgin Galactic, Iridium, Maxar and Viasat, which are involved mainly in satellite-based services, in its portfolio.[16] These two ETFs hint that the interplanetary connectivity is about to become a bullish economic undertaking with precise capital gains.

With such financial mechanisms in play, astropolitical blocs could maintain protectionist barricades, and osmosis, if any, will be restricted to mergers and acquisition of technologies, or companies, or human resources. With technologies and finances in place, astropolitical blocs intend to institutionalise codes of conduct conducive to their end goals, restrict profits within the blocs, delimit sharing of gains with groups outside their bloc and patrol and restrict access and benefits of other blocs. Astropolitical blocs thus cooperate within the bloc and promote contestation with those outside it.

Notes

1 M. Young, "Bad Idea: The Wolf Amendment (Limiting Collaboration with China in Space," *Defence 360° - Center for Strategic and International Studies*, 4 December 2019, https://defense360.csis.org/bad-idea-the-wolf-amendment-limiting-collaboration-with-china-in-space/

2 International Space Exploration Coordination Group website, https://www.globalspaceexploration.org/

3 "Global Exploration Roadmap Supplement – Lunar Surface Exploration Scenario Update," *International Space Exploration Coordination Group*, 28 August 2020, https://www.globalspaceexploration.org/?p=1049

4 "JAXA and Toyota Announce "LUNAR CRUISER" As Nickname for Manned Pressurized Rover," *Japan Aerospace Exploration Agency and Toyota Motor Corporation*, 28 August 2020, https://global.toyota/en/news-room/corporate/33208872.html

5 "Battelle Energy Alliance seeks industry partners to design nuclear power system for moon," *Idaho National Laboratory News Release*, 24 July 2020, https://inl.gov/article/battelle-seeks-partners-moon-technologies/

6 "Nokia selected by NASA to build first ever cellular network on the Moon," *Nokia Communications*, 19 October 2020, https://www.nokia.com/about-us/news/releases/2020/10/19/nokia-selected-by-nasa-to-build-first-ever-cellular-network-on-the-moon/

7 F. Weiwei, Y. Fan, H. Lin, W. Haiming, "Overview of Russia's future plan of lunar exploration," *Science & Technology Review* 37, (2019), 6–11.

8 "About NEK & SIRIUS – Mission Overview," https://www.nasa.gov/analogs/nek/about

9 Z. Lei, "Senior officer expects moon visit by 2036," *China Daily*, 29 April 2016, https://www.chinadaily.com.cn/china/2016-04/29/content_24957196.htm

10 "2019 Report to Congress of the U.S.-China Economic and Security Review Commission," U.S.-China Economic and Security Review Commission, November 2019, https://www.uscc.gov/sites/default/files/2019-11/2019%20Annual%20Report%20to%20Congress.pdf

11 "United Nations/China Cooperation on the Utilization of the China Space Station," *United Nations Office for Outer Space Affairs and China Manned Space Agency*, 2018, https://www.unoosa.org/documents/doc/psa/hsti/CSS_1stAO/Leaflet_18-05889_Cooperation_ebook_updated14092018.pdf

12 "United Nations/China Cooperation on the Utilization of the China Space Station – Selected Experiment Projects to be executed on board the CSS for the 1st Cycle," *United Nations Office for Outer Space Affairs and China Manned Space Agency*, 12 June 2019, https://www.unoosa.org/documents/doc/psa/hsti/CSS_1stAO/1stAO_FinSelResults.pdf

13 "SOM Releases Concept for Moon Village, the First Permanent Human Settlement on the Lunar Surface," Press Release - Skidmore, Owings & Merrill LLP, 9 April 2019, https://www.som.com/news/som_releases_concept_for_moon_village_the_first_permanent_human_settlement_on_the_lunar_surface

14 "China Proposes Establishing Moon-Based Special Economic Zone," China Briefing – Dezan Shira & Associates, 8 November 2019, https://www.china-briefing.com/news/china-proposes-establishing-moon-based-special-economic-zone/

15 From State Street Global Advisors website, https://www.ssga.com/us/en/institutional/etfs/funds/spdr-sp-kensho-final-frontiers-etf-rokt

16 From the PROCUREAM website, https://procureetfs.com/etfs/ufo.html

5 Plugging India into interplanetary connectivity projects

Chapter 2 of this book chronicled the emergence of India's space exploration programme briefly. The recount immediately points at India's diplomatically amicable space programme with immense respect for nations with superior space capabilities and amicability with those in lower ranks of competencies.

India's civilian space programme is well entrenched into numerous international space science, technical and legal multilateral bodies. It is a founding member in many of them. It actively participates in the United Nations Committee on the Peaceful Uses of Outer Space (UN-COPUOS), COSPAR, IAF, International Academy of Astronautics (IAA), Space Frequency Coordination Group (SFCG), Coordinating Group for Meteorological Satellites (CGMS) and the Committee on Earth Observing Satellites (CEOS).

ISRO recently finished its term as the Chair of the CEOS, an inter-governmental body established within the G7 Economic Summit of Industrial Nations Working Group on Growth, Technology and Employment, for 2020.[1] True to its temper for regional cooperation, India's priority areas for the CEOS in that year were:

- analysing gaps in priority areas of satellite-based measurements such as winds, forest cover, emissions and disaster situational awareness,
- use of space-based applications to ensure execution of United Nations Sustainable Development Goals (SDGs) in the regional Bay of Bengal Initiative for Multi-Sectoral Technical and Economic Cooperation (BIMSTEC) multilateral,
- assessment of renewable energy potential (wind, solar energy) and disaster prediction and monitoring through space-based applications.

DOI: 10.4324/9781003152934-5

On similar lines, the Indian space policymakers and technocrats can bring a cooperative sentiment in working groups and multilateral bodies increasingly focusing on planetary exploration.

India's posture in the planetary exploration multilaterals

In the past few years, numerous space multilaterals have formed new working groups and action teams dedicated to the space exploration domain. Their formation is attributed chiefly to the concerted progress characteristic of the Second Space Age.

For instance, on the United Nations Conference's Fiftieth Anniversary on the Exploration and Peaceful Uses of Outer Space (UNISPACE+50), the UN-COPUOS identified priority areas and intergovernmental cooperation mechanism relevant to the coming times in space exploration.[2] The priority areas include:

* Form global partnership in space exploration and innovation to address global challenges, promote cooperation between spacefaring nations and foster dialogue with the private space sector.
* Promote UN treaties' universality on outer space activities. Address possible gaps in outer space legal regimes. Foster an international regime of liability and responsibility to mitigate challenges to security, safety and sustainability of outer space activities.
* Enhance the exchange of information on space-based objects and events for risk-reduction notification and confidence-building mechanisms between member nations.
* Recognise space weather as a global challenge. Develop international coordination and exchange mechanisms to mitigate extreme space weather impact and develop risk analysis and assessment protocols.

Being a signatory to all major UN space treaties on outer space, India has proactively taken steps to build capacities in this direction. The space reforms of May 2020 foster a national dialogue with the private sector and goes one step ahead by giving them an equal pedestal. The drafting of a new comprehensive space policy could be a step towards forming a national regime for deep space activities. The establishment of ISRO's Space Situational Awareness Control Centre near Bengaluru and ISRO's Project Netra are strides in developing space-debris tracking abilities.[3] The Department of Space, the Department of Science and Technology and the Ministry of Education have space weather research and monitoring competencies that are growing.[4] All these leads hint to the extent of India's preparedness for meeting with the various priority areas, connected indirectly with the interplanetary connectivity goals laid down by UN-COPUOS.

India's planetary missions are vital cogs in the Global Exploration Roadmap. India as a team-player apprises its ISECG partners with its space exploration mission designs. The ISECG member space agencies offer housing for foreign payloads on their interplanetary spacecraft and share spacecraft housekeeping and payload science datasets. The ISECG coordinates the types of measurement payloads each member can use for a particular type of space mission. The technical specifications of the payloads differ, thereby offering a more comprehensive set of scientific measurements. India's scientific institutions, university-based laboratories and DoS laboratories often participate in multilateral scientific and technical platforms like the IAF, IAU and COSPAR to exchange scientific know-how. Apart from ISECG, most of the multilateral bodies mentioned above are around five decades old and have become too rigid and focused on accommodating the new facets of planetary exploration in the emerging space economy.

In 2020, amid the disruption caused by the COVID-19 pandemic, the UNOOSA launched the 'Space Economy Initiative' under its ambits. The initiative aims to bring non-space faring and spacefaring nations on one platform and identify methods to strengthen their space economies.[5] This initiative aims to enhance cooperation between its member states' public and private sectors and promote the global space economy's inclusive and sustainable progress.

UNOOSA took another critical step by organising a Space Economy Leaders Meeting during Saudi Arabia's Presidency of G20 multilateral in 2020. If the meeting is to attain equivalence with other G20 sub-meetings such as Business 20 (B20), Think Tank 20 (T20) and Science 20 (S20), it will become an annual fixture in this crucial grouping. The G20 is among the most potent non-UN multilateral forum in the world. The countries amount to almost 90% of the gross world product, almost 80% of the total world trade volume and two-thirds of the world population. A frequent discussion on the space economy in Space20 may churn new financing mechanisms, trade, trade in services regimes, bilateral, minilateral and multilateral collaborations and regulations for new and evolving space-businesses.

India's space reforms of May 2020 were announced cumulatively by the Prime Minister's Office and the Union Cabinet through Minister of Finance Nirmala Sitharaman.[6] The pronouncement of the details of the space reforms by the Finance Minister was not unintentional. The reforms were considered fast-moving developments assessed by the steps taken up by UNOOSA in Vienna and later during Saudi Arabia's G20 presidency. Numerous business consulting firms, like Morgan Stanley, KPMG and McKinsey, with a strong presence in India and industry bodies like Confederation of Indian Industries (CII) and Federation of Indian Chambers of Commerce and Industry (FICCI), have been making noise about the bullish investments in the space sector and the steady growth of the global space economy. That aside, the grand yet pragmatic steps taken by new and emerging spacefaring countries have also impacted New Delhi. The space reforms of May 2020 are a sign of India's self-reliant posture.

Suryamala: Aiming for multi-modal-like connectivity megaproject

India is building an enormous multimodal transportation infrastructure to enhance road, rail, air and sea-based connectivity. Infrastructure is a crucial driver of India's economy. Therefore the country's gross fixed capital formation has remained in fine fettle between 30% and 35% over the past few years.[7] India is banking heavily on infrastructure and digital connectivity sectors for its economic growth in the coming decades.[8] Both sectors are linked intimately to the space economy. Space-based communication, navigation, surveillance and reconnaissance systems are vital for terrestrial connectivity of any kind.

Today, *Bharatmala* (road connectivity),[9] *Diamond Quadrilateral* (rail connectivity),[10] *Sagarmala* (inland and coastal port-based connectivity)[11] and *UDAN-Regional Connectivity Scheme* (air connectivity)[12] are considered as important arteries of India's economy. India's services, agriculture, energy and manufacturing sectors are all dependent on the strengthening of connectivity infrastructure. Suppose India is to contemplate interplanetary connectivity, similar to its four terrestrial connectivity projects, in that case, its massive socio-economic advantages will become evident.

India can converge its *Chandrayaan*, *Mangalyaan* and *Gaganyaan* trio as part of *Suryamala*. *Suryamala* could take cues from the terrestrial connectivity megaprojects. It can emphasise employing the public and private sectors to innovate, construct, launch, operate and provide second-order services and manufacturing via suitable project delivery methods.

The five introductory infrastructure provisions of the global Earth-Moon-Mars interplanetary connectivity plans that *Suryamala* can pursue are:

Communications: Raise, operate and serve high-fidelity, low-latency interplanetary and cis-lunar (Earth-Moon) telecommunications connectivity with indigenous technologies with select international partners;

Logistics: Develop propulsion and spacecraft technology for building and maintaining cost-effective interplanetary transportation services supply chains;

Surveying: Partake in international scientific planetary surveying and prospecting missions for Moon and Mars to identify locations suitable for human habitation and to set up robotic stations;

Environmental Sustainability: Build space debris mitigation, effluent recycling and life-support technologies for operating in extraterrestrial environments.

Energy: Develop clean fuel and solar technologies for long-duration space missions.

Suryamala could become a key driver for raising India's share in the global space economy. As India's space innovation ecosystems entrench in *Suryamala*, they will realise numerous space technologies' spin-offs for various terrestrial applications. *Suryamala* can contribute to the high-technology manufacturing goals set by the Indian government under the Make in India programme. On the international affairs front, *Suryamala* could be fitted appropriately with the Global Exploration Roadmap of the ISECG, with UNOOSA's Space Economy Initiative and the tentative motivations of the Space20 grouping.

A national space vision for interplanetary connectivity

The task of setting up an interplanetary connectivity infrastructure is not a one-off space mission of anyone's space agency. It needs delegation to all the countries participating in the global space economy and consistent leadership from a select few economically, socially, scientifically and militarily secure nations. Therefore, it is evident that a select few superpowers are interpreting their version of interplanetary connectivity, be it through the US' Artemis Accords, Europe's Moon

Village or China's Earth-Moon Special Economic Zone. Interplanetary connectivity demands a non-partisan national vision, consistent budgetary commitments, robust execution plans, steady profit flows and stringent environmental norms and protocols. *Suryamala* could become a meta-strategic megaproject for India.

The global space economy's projected growth is known to academia, consulting firms, industry bodies, think tanks and governmental echelons. However, setting a national space vision for interplanetary connectivity requires careful political crafting and narrative setting. India has a political history of deracinating futuristic space activities and deeming them too fantastic for a developing country. India's space exploration programme suffered from such political contempt. It was hence commenced 30 years after the commencement of ISRO. Unlike space exploration which was entirely a scientific activity, interplanetary connectivity being socio-economic cannot be reasoned as excessive engagement. However, there is a need to set a narrative to deem interplanetary connectivity as pragmatic and feasible.

To that end, India must chalk out a national space vision for the rest of the 21st century. There is a precedent to such long-range planning in terms of the space economy. In 2017, the UAE, an emerging spacefaring nation, set a target of establishing a city on planet Mars by 2117. In this period of 100 years, the UAE has set its eyes on attracting the world's best talent that can achieve this target for it. In January 2021, Abu Dhabi announced offering citizenship to investors, specialised professionals, intellectuals and innovators.[13]

New Delhi has often shied away from making long-term plans. It stuck with the custom of short-term and myopic five-year planning ingrained by the erstwhile Soviet-style governance. The five-year plans often cite structural uncertainties, black swan events and risks of over-pessimism and over-optimism as impediments for long-range planning. Long-range planning is arduous in a geopolitically multipolar world. On the plus side, the Second Space Age will make it imperative for nations concerned about their power status to give due diligence for long-range planning.

Interplanetary activities span over months and a few years. For example, a mission to Mars requires six months of travel in one direction, and a simplistic sample-return mission will require no less than a year. A short-term human stay on Mars will demand meticulous planning in life-support systems, ration, electric, electronic, communications, energy and other systems onboard the spacecraft and the enclosed habitat. A year-long mission will require rigorous planning, research and development on the scale of thousands of human hours.

Long-range planning and forecasting are essential tools of strategic management. This 'holy grail' of a tool is absorbed only by nations who aspire to build or maintain their superpower status. Therefore, India's top leadership must beget a national space vision at least until the year 2100. A national space vision will offer clarity to the diverse technocratic stakeholders of *Suryamala* and India's other commercial, military and civilian space activities.

Notes

1 Retrieved from the *Committee on the Earth Observing Satellites* website, https://ceos.org/about-ceos/organization/
2 Committee on the Peaceful Uses of Outer Space, "Draft resolution entitled "Fiftieth anniversary of the first United Nations Conference on the Exploration and Peaceful Uses of Outer Space: space as a driver of sustainable development," *United Nations General Assembly – Vienna*, 20–29 June 2018, https://www.unoosa.org/res/oosadoc/data/documents/2018/aac_105l/aac_105l_313_0_html/V1803310.pdf
3 "ISRO SSAControl Centre Inaugurated by Dr. K. Sivan, Chairman, ISRO/Secretary, DOS," *Indian Space Research Organization*, 16 December 2020, https://www.isro.gov.in/update/16-dec-2020/isro-ssacontrol-centre-inaugurated-dr-k-sivan-chairman-isro-secretary-dos#:~:text=NEtwork%20for%20space%20object%20TRacking,ISTRAC%20campus%20at%20Peenya%2C%20Bangalore
4 A. Bhardwaj, T.K. Pant, R.K. Choudhary, D. Nandy, & P.K. Manoharan, "Space weather research: Indian perspective," *Space Weather* 14, (2016), 1082–1094. https://agupubs.onlinelibrary.wiley.com/doi/full/10.1002/2016SW001521
5 United Nations Office for Outer Space Affairs, "Space Economy Initiative – 2020 Outcome Report," January 2021, https://www.unoosa.org/documents/pdf/Space%20Economy/Space_Economy_Initiative_2020_Outcome_Report_Jan_2021.pdf
6 Press Information Bureau - Ministry of Finance, "Finance Minister announces new horizons of growth; structural reforms across Eight Sectors paving way for Aatma Nirbhar Bharat," Government of India, 16 May 2020, https://pib.gov.in/PressReleseDetail.aspx?PRID=1624536
7 A. Dangra, "The Missing Piece in India's Economic Growth Story: Robust Infrastructure," *S&P Global*, 2 August 2016, https://www.spglobal.com/en/research-insights/articles/the-missing-piece-in-indias-economic-growth-story-robust-infrastructure
8 "Indian Infrastructure Sector in India Industry Report," *India Brand Equity Foundation*, January 2021, https://www.ibef.org/industry/infrastructure-sector-india.aspx
9 Press Information Bureau – Ministry of Road Transport & Highways, "Progress of Bharatmala Project," *Government of India*, 9 December 2019, https://pib.gov.in/PressReleseDetailm.aspx?PRID=1595501
10 Press Information Bureau – Ministry of Railways, "Diamond Quadrilateral Bullet Train Network Project," *Government of India*, 29 November 2019, https://pib.gov.in/newsite/PrintRelease.aspx?relid=195193

11 Press Information Bureau – Ministry of Ports, Shipping and Waterways, "Sagarmala Programme," *Government of India*, 12 March 2020, https://pib.gov.in/PressReleasePage.aspx?PRID=1606138
12 Press Information Bureau – Ministry of Civil Aviation, "78 New Routes Approved Under UDAN 4.0," *Government of India*, 27 August 2020, https://www.pib.gov.in/PressReleasePage.aspx?PRID=1648927
13 "With citizenship world's talent can call the UAE home," *The National Editorial*, 1 February 2021, https://www.thenationalnews.com/opinion/editorial/with-citizenship-world-s-talent-can-call-the-uae-home-1.1156886

6 The economics of interplanetary connectivity

During the First Space Age, the narrative for space exploration revolved around the scientific and geopolitical one-upmanship between capitalism and communism. The narrative has become more nuanced in the Second Space Age owing to the now multipolar global geopolitics. It is driven by the contest for the global space economy, investments, infrastructure and connectivity. Due to this multipolarity, many nations aim to control substantial stakes in the new business-oriented perspective of space exploration. Their stakes depend on their ability to build, operate and sustain an interplanetary connectivity infrastructure unaided or as part of astropolitical blocs.

Scientific RDTE will continue to be the undercurrent of space exploration. Nevertheless, its greater purpose is shifting towards setting economic targets and attaining metastrategic national goals. This shift is also metamorphosing the once 'space sector' into a full-fledged 'space economy.' Outer space is inaccessible and hostile to both machines and humans. So, the farther they travel and operate in outer space, space activities will become less idealistic and more rational, laden with *realpolitik*.

Innovation and resource-driven technocratic space industrial complex

Nations worldwide have a near-unanimous comprehension that the bedrock of nations' socio-economic and military strength lies in possessing multifarious space capabilities. There is no superpower in the world that has shunned space capabilities deeming it unworthy. The global space economy has grown by approximately 10 billion dollars annually since 2016. It was estimated to be reaching $423.8 billion in 2020.[1] Furthermore, by 2040, conservative estimates suggest the global space economy will touch US$ 3 trillion.[2] The commercial satellite

DOI: 10.4324/9781003152934-6

industry has the largest share currently in the global space economy. The coming two decades could experience exponential growth of the global space economy through innovations in satellite and non-satellite industries. Interplanetary connectivity and exploration can become a cornerstone for the non-satellite sector.

The US Artemis programme has scheduled lunar exploration missions during the entire decade of 2020s. The lunar landers, lunar rovers, crewed capsules, launch vehicles and other components, part of the Artemis programme, are sourced mainly from commercial space contractors. These private contractors possess various specialisations: space robotics, autonomous industrial systems, spacecraft modelling and simulations, aerospace manufacturing, specialised payload manufacturers, semiconductors, space-grade electronics, software, launch vehicles, spacecraft manufacturing and many more. By employing them all, the Artemis programme has tapped into the vast innovation and manufacturing potential within the US, generated contracts for these industries to sustain and converted the national-budget-dependent space exploration activity into a flourishing 'lunar' economic sector.

Regardless of Artemis's progress and eventuality, Europe, China, Japan, India and few other countries are modeling their respective space exploration aspirations based on Artemis. Artemis and its predecessor, Commercial Lunar Payload Services (CLPS), CCDev, CRS and COTS have off-loaded the extensive functions of governmental space agencies and distributed them to private industries. Space agencies of many democratic countries are no more interested in being the end-to-end manufacturer and service providers. They now have privatised large sections of commercially viable space services and manufacturing. Space agencies are focusing on being a facilitator-regulator and taking up only commercially unviable high-risk, high-reward space projects. Dispensing opportunities to the private sector has allowed the latter to innovate, identify cross-sectoral spin-offs, commercialise them and build market heft. With greater brand valuations and strong perception management revolving around many private space companies, getting recruited by them has become an aspiration for the potential workforce. National space agencies are turning into high-risk, high-reward incubators and facilitators for space sector startups. The startups have become incubatees and contractors, with scientific research institutions as sources of their ideas, concepts and prototypes. The interplanetary connectivity success will depend on such ecosystems' functioning.

Interplanetary connectivity megaprojects are prone to numerous success-limiting factors such as technical glitches, human-introduced errors, financial risks and mismanagements and accidents. These inhibitory factors will demand efficient insurance services, technology standards bodies, technical inspection, testing, certification bodies, disaster recovery and search and rescue services. Moreover, no one nation, howsoever economically resilient, a strong talent magnet and space capable, is eager to swallow all these success limiting factors unaccompanied. To this end, they are forming astropolitical blocs. They are ready to plug in their domestic space industrial complex with the counterpart ecosystems in their partner countries.

This trend of sharing resources by creating supra-national space industrial complexes has begun to extend, as is visible from the formation of the astropolitical blocs. Moreover, these blocs are expanding in terms of the partner countries joining in. However, like in every ecosystem, there is an apex predator. Space industrial complex extending across astropolitical blocs has a lead country that concentrates most of the space capabilities compared to the others. However, what binds them are ideological and cultural similarities, a geopolitical assemblage and common strategic interests. Therefore, no country, advanced or developing, intends to shun space activities.

Revisiting the notion of supra-global commons

The notion of deeming outer space and all the celestial bodies, planets, satellites and Small Solar System bodies as global commons began immediately out of the geopolitical Space Race between the US and the Soviet Union. Ten years after the launch of Sputnik, in 1967, the United Nations got the Outer Space Treaty ratified with maximum support from all over the world, including the two countries engaged in the space race. For decades, the Outer Space Treaty and its framework were untouched, given the fact that there were only two players who could reach the Moon and other planets. Those space agencies that later acquired the capabilities, particularly the ESA and JAXA, never intended to perturb the treaty's legalities.

Nevertheless, with increasing geopolitical multipolarity, each astropolitical bloc wants to cultivate exclusive interplanetary connectivity capacities and end goals. Therefore, the idea of global commons is on the precipice of drastic reviews. Global commons' drastic reviews are apparent in democratic systems than autocratic ones because democratic nations publicise national policies and legislation. In contrast, autocratic nations do away with legislation as their supreme leaders' assertions hold higher esteem than votes in democracies.

The Science and Technology Commission of the China Aerospace Science and Technology Corporation has recently announced its intention to establish an Earth-Moon Special Economic Zone. It has estimated an investment of around 10 trillion dollars in the same by the year 2050. The Earth-Moon Special Economic Zone consists of affordable transport and power systems, data connectivity and lunar surface presence. The CLEP is perhaps the second most successful lunar exploration programme after Apollo. It has realised five strategically successful – Chang'e – missions in around a decade. China's proven ability to land and relay signals from the lunar far side[3] demonstrates its avowal to lodge early presence on the Moon. Early arrivers on the Moon are likely to determine the rules of behaviour and permit access for others. This assertive competence and intention are entirely outside the aging contest's ambits between the US and Russia. China's activities on the Moon and its intent to lead its astropolitical bloc is the most prominent signal of multipolar geopolitics transforming into multipolar astropolitics.

As evident from the Artemis Accords, the US pre-empted the fast-paced capability build-up across astropolitical blocs, a US-led astropolitical bloc with Canada, the United Kingdom, Luxembourg, Japan, Australia, New Zealand and the United Arab Emirates as member countries. In 2013, an Apollo Lunar Landing Legacy Bill was put forth in the US Congress. The declined bill proposed to empower NASA to manage Apollo landing sites on the Moon as designated US national parks. The proposition of such a colonialist legislation has set precedence for other autocrat countries to follow suit. It is certain, in the 2020s and 2030s, numerous international space treaties will undergo revisions: *de jure* and *de facto*. However, such revisions will see intense debates so that they do not damage national interests.

It will be difficult for any nation by astropolitical blocs to lay sovereign claims over vast lunar, Martian or other extraterrestrial real estates. In the next two decades, astropolitical blocs will set up competitive space transportation systems, cis-lunar and interplanetary telecommunication networks, and robotic and crewed long-duration research stations on lunar and Martian surfaces. These blocs will compete over technological competencies and maintain starkly independent and incompatible technology standards. The entire connectivity infrastructure remains exclusive to the users from the blocs.

Global commons' notion and framework will experience tremors in the next two decades. The tremors' causal factors will be the expansive and simultaneous progress made by the astropolitical blocs in building their independent interplanetary connectivity infrastructure. Setting up multiple interplanetary connectivity infrastructure is coherent and relevant to the world's techno-economic progress. Sustainable global economic progress has intimate links with space-bound investments.

Sustaining interplanetary connectivity

Establishing interplanetary connectivity and sustaining it is a high-priced and long-duration commitment. It demands sustained political patronage, robust planning and management, consistent economic returns and strong international partnerships. These factors are variable, making it less likely for middle- and low-income economies to pick up large stakes in the astropolitical blocs they belong. Continuing to associate with a bloc can help middle- and low-income economies build niche technology capabilities. With such capabilities, some of them could eventually become indispensable to global space economy value chains. Becoming indispensable to value chains can help their broader industrial base, attract talent with exciting deals and improve their national socio-economic indicators.

Astropolitical blocs will make a case for the equitable development of all the bloc partners but under a bloc leader's pre-eminence. The US, China and Russia will aim to be bloc leaders. Countries like India, France and Germany may not align with any of the astropolitical blocs to upkeep their strategic autonomy. The bloc constitution could remain dynamic with new partners joining in while some partners could jump blocs or become neutral.

Space competent neutral nations will have inherent advantages despite missing out on the securities that bloc partners will enjoy. They will collaborate across multiple blocs as their niche proficiencies will be of value to them. They will acquire expertise, absorb best practices and attain crucial and nonpareil standing in the global space economy but lesser than the astropolitical bloc leaders. However, the most significant advantage of the neutral nations will be their ability to assuage tensions and conflicts between the blocs and become essential arbitrators in rapprochement mechanisms.

Sustaining interplanetary connectivity also demands new global regulations and liability clauses for damages caused to space-based assets and properties and damages to the global commons. Sustainable interplanetary connectivity will also depend on conflict-resolution and confidence-building measures between astropolitical blocs. Patrolling agencies, of the likes of International Criminal Police Organization

(INTERPOL), the Preparatory Commission for the Comprehensive Nuclear-Test-Ban Treaty Organization (CTBTO), International Search and Rescue Advisory Group (INSARAG), among others, is needed. An INTERPOL-like body will be needed to keep a check on criminal activities in outer space. A CTBTO modelled body could monitor the illegal usage of weapons of mass destruction. An INSARAG-like body could assist with the search and rescue of lost assets and crew in distress.

Interplanetary connectivity will increase the demand for clean fuels and critical minerals. These astropolitical blocs will attempt to secure the demands using the blocs' energy and critical mineral supply chains. So, every bloc will have one or more partner countries abundant with the natural reserves of necessary fuel and minerals. The global impetus on hydrogen and nuclear energy will find a tremendous boost due to its applicability for space transportation. The extraction of hydrogen and oxygen from the lunar and Martian regolith will make it an attractive cryogenic fuel for surface launches. The space launch industry already realises this potential. Most of them are working around cryogenic fuels and reducing the impetus on conventional rocket fuels.[4]

Sustaining interplanetary connectivity is a multifarious global undertaking. It needs both competitions and cooperation between the astropolitical blocs and within them. Therefore discussing cooperation and competition on multilateral track-one diplomatic fora, such as the G20, Group of Seven (G7) and the Permanent Five of the Security Council, becomes extremely important. Most members and observer countries in these fora are members of one or the other developing astropolitical blocs, including non-aligned. A track-1 diplomatic engagement can dissipate disputes and serve as a geopolitical hotline for any burgeoning conflict.

All these bodies will contribute to maintaining the core philosophy of global commons. The technological and socio-economic progress they have taken charge of will make amendments to the global commons clauses that inhibit this progress. Consensual revisions in the understanding of global commons will be central to sustaining interplanetary connectivity.

Notes

1 R. Cooper, "Global space economy grows in 2019 to $423.8 Billion, The Space Report 2020 Q2 Analysis shows," *The Space Foundation*, https://www.spacefoundation.org/2020/07/30/global-space-economy-grows-in-2019-to-423-8-billion-the-space-report-2020-q2-analysis-shows/#:~:text=Global%20Space-,Global%20Space%20Economy%20Grows%20in%202019%20to%20%24423.8%20Billion%2C%20The,Report%20

2020%20Q2%20Analysis%20Shows&text=COLORADO%20
SPRINGS%2C%20Colo.&text=Space%20products%20and%20
services%2C%20a,1.7%25%20from%202018's%20%24214.18%20billion

2 K.W. Crane, E. Linck, B. Lal, & R.Y. Lei, "Measuring the space economy estimating the value of economic activities in and for space," *Science and Technology Policy Institute, Institute of Defense Analyses*, March 2020, https://www.ida.org/-/media/feature/publications/m/me/measuring-the-space-economy-estimating-the-value-of-economic-activities-in-and-for-space/d-10814.ashx

3 D. Castelvecchi, "China becomes first nation to land on the Moon's far side," *Nature*, 3 January 2019, https://www.nature.com/articles/d41586-018-07796-x

4 E. Stoll, P. Härke, S. Linke, F. Heeg, & S. May, "The regolith rocket – A hybrid rocket using lunar resources," *Acta Astronautica* 179, (2021), 509–518, https://www.sciencedirect.com/science/article/abs/pii/S0094576520306962

7   The second half of the 21st century
Will humans become interplanetary species?

Humans becoming an interplanetary species is perhaps one of the most celebrated aspirations of many cultures and nations. The First Space Age saw the earliest steps taken towards that goal, and since then, many national space programmes have contributed to it in kind.

Humans' persistence to go interplanetary is driven by the curiosity and dexterity innate to a highly intelligent species and the existential risks to life on Earth from natural and anthropogenic extinction-level events and chronic planetary threats. For a long time, futurists have recognised residing on one planet as a significant security concern for humanity's sustenance and survival. Global pandemics, weapons of mass destruction, geological disasters, extreme weather events, bolide impacts, chronic issues like climate change, population growth pressures and depletion of natural environmental processes and essential but limited and non-renewable resources are significant existential extinction-level risks for humanity.[1] Major spacefaring nations have begun to invest consistent efforts to deduce short-term and justifiable rationale to move to other planets perhaps by the end of the 21st century. Doing so, they aspire to expand the global economy in near-Earth space by developing interplanetary connectivity networks.

Although shared commonly by all astropolitical blocs, this motivation has onto it impressions of the ongoing international geopolitics. It, therefore, becomes apparent that humanity will not go into outer space as a monolithic unit. Many nations will pursue it in an organised manner through astropolitical blocs with like-minded partners.

The extraterritoriality of nations

The 1967 Outer Space Treaty restrains space-capable nations from making sovereign exclaves in outer space and on the Moon and Mars surfaces. The prohibitory Article II states[2]:

DOI: 10.4324/9781003152934-7

Outer space, including the moon and other celestial bodies, is not subject to national appropriation by claim of sovereignty, by means of use or occupation, or by any other means.

Article II has endured since its inception, as no nation has attempted to form sovereign territories on an extraterrestrial body. Even in the present multipolar astropolitical circumstances, with precise positions of bloc leaders like the US, Russia and China, none are able to singularly exert territorial claims on extraterrestrial surfaces. One such attempt was made when the 2013 Apollo Lunar Landing Legacy Bill was tabled in the US Congress.[3] The bill intended to deem the Apollo lunar landing sites like national parks of the US, indirectly making it a national appropriation case. The bill also went ahead to define Apollo mission equipment lying on the Moon as artefacts, which are to be preserved and protected for posterity. This bill was declined immediately as it was conflicting with Article II. However the latter point of artefacts, which does not conflict with Article II, can become a yardstick for newer extraterritorial intentions of the US and perhaps even other astropolitical bloc leaders.

Astropolitical bloc leaders are aware that despite their technological prowess and economic heft, their national claims too are legally contestable. But what is considerably difficult for them is the sustenance of the entitlements they seek both economically and technologically. As a potential way out, the bloc leaders are taking along their like-minded partners in the astropolitical blocs to foster collective assets in outer space. Since Article II does not have restrictions on appropriation by blocs, astropolitical bloc leaders could position assets to acquire extraterritorial and extraterrestrial real estate in unison with their bloc partners.

The draft Space2030 Agenda,[4] a recent undertaking of the UN-COPUOS, has expanded the scope of consensus between UN member states that now includes space accessibility, space economy, space diplomacy and space society. The agenda encourages member states to actively inculcate bilateral, multilateral, regional and international space cooperation. The openness of the UNOOSA accompanies the encouragement to receive extra-budgetary donations to advance the execution of the Space2030 Agenda. These developments suggest that astropolitical blocs will identify themselves as multilateral bodies and make their scientific and economic activities legally tenable. Within an astropolitical bloc, nations may identify their space-based assets in the interplanetary connectivity projects as national properties chartered to multilateral bodies. With that reasoning, the

Chinese Large Modular Space Station, expected to be raised from 2021 onwards, will be a property of the People's Republic of China. Likewise, the infrastructure that constitutes the Moon Village will become the property of the member states of ESA and partners. Similarly, the assets deployed by the Artemis Accords partners will be owned and controlled entirely by them.

The more significant number of assets a particular country has on extraterrestrial surfaces, the more significant their claim over the regions where these assets land. Since astropolitical blocs will primarily undertake interplanetary connectivity and exploration, only the bloc partners will manage the assets and the locations where they are grounded.

New states, new rules, new extraterritorial order

The prospective interplanetary connectivity infrastructure between the Earth, Moon, near-Earth asteroids and Mars could be owned by private and national corporations with national governments' stakes. This public–private infrastructure in the global commons will demand a new code of conduct, terms of references, rules and treaties. Such regulations will emerge from the little consensus between different astropolitical blocs. These blocs will delineate access to countries outside the blocs based on their strategic favourability.

There are also chances that countries that uphold the doctrine of strategic autonomy, non-aligned to the blocs and have armed neutrality will play a significant role in shaping the norms of interplanetary connectivity projects. These countries may form loosely bound groupings even with the members of astropolitical blocs based entirely on short-term economic interests. Such purely economic interests will cause their space agencies to corporatise such that these agencies set goals on profits along with scientific innovation and discoveries. The corporate headquarters-like lean operations of space agencies will also stem from the ongoing commercialisation of space activities and the private sector's increasing role in building interplanetary connectivity in *terra nullius*.

The legal tenets of the Outer Space Treaty will not suffice in the economics-driven Second Space Age. Many of the space activities will be governed by corporate, competition, consumer, public service, insurance, insolvency and environmental laws and by the rights and duties of investors, shareholders, stakeholders, directors, promoters and consumers. Interplanetary connectivity projects will set off an immense potential for international regulatory bodies. The bodies will monitor best business practices in space environments, various types of due diligence, compliance to international treaties and code of conduct, auditing and risk assessment maintaining robust testing and verification of technologies and processes, and upkeeping software and hardware technology standards blocs and globally. The space law and regulation domains is about to expand in its scope tremendously.

The nations in charge of astropolitical blocs and those that have heft over intergovernmental bodies in charge of maintaining law and order in outer space, as it is comprehended presently, will be the ones defining new rules and determine astropolitical order conducive to them. The greater the techno-economic capabilities and heft, the higher are the chances of that nation exercising its order in interplanetary space. The bloc partners will align with a leader to benefit economically and be a beneficiary of the astropolitical order that emerges from their joint interplanetary economic activities.

Biological concerns from interplanetary connectivity in the post-COVID era

The COVID-19 global pandemic has reiterated the significance of greater awareness about global biological and chemical security concerns. In the post-COVID-19 era, controlling the deliberate and non-deliberate transport, from Earth-to-space and space-to-Earth, of hazardous and unverified biological and chemical materials will become highly crucial. These threats will only escalate if the interplanetary connectivity infrastructure is built on lax 'planetary protection' protocols and quarantine measures and fail to abide by the Article IX of the Outer Space Treaty, which states:

> States Parties to the Treaty shall pursue studies of outer space, including the Moon and other celestial bodies, and conduct exploration of them so as to avoid their harmful contamination and also adverse changes in the environment of the Earth resulting from the introduction of extraterrestrial matter, and where necessary, shall adopt appropriate measures for this purpose.

The international space science and research organisation, COSPAR, has identified itself with investigations of forward (Earth to space) and backward (space to Earth) biochemical contamination and human spaceflight since its initiation.[5] The COSPAR categorises space missions into five categories (Table 7.1).

Table 7.1 Categories of planetary protection formulated by the Committee on Space Research

COSPAR planetary protection categories of space missions	Current description of categories	Criterion in the category	Extent of planetary protection (scale of very low to very high)
Category I	Mission to a target celestial body not aiming to understand the process of chemical evolution or the origin of life	Mission/infrastructure should be sanitised and placed in locations on celestial bodies with low or no presence of native extraterrestrial biochemical signatures	Very low
Category II	Missions to target celestial body with significant interest in processes pertaining to chemical evolution or origin of life	Spacecraft/infrastructure should be sanitised despite remote chance of contamination of locations of biochemical interest	Low
Category III	Missions to target celestial body with high interest in processes pertaining to chemical evolution or origin of life	Criteria of Category II + Need for trajectory biasing, use of cleanrooms for spacecraft/infrastructure assembly and bioburden reduction	Medium

(*Continued*)

Table 7.1 (*Continued*)

COSPAR planetary protection categories of space missions	Current description of categories	Criterion in the category	Extent of planetary protection (scale of very low to very high)
Category IV	Missions to target celestial body with assured signatures of processes pertaining to chemical evolution or origin of life	Criteria of Category III + use of bioassays for bioburden reduction, biochemical contamination analysis, bulk organic analysis, partial sterilisation of spacecraft and infrastructure hardware and complete sterilisation of direct-contact hardware	High
Category V	All Earth-return missions from celestial bodies with assured signatures of processes pertaining to chemical evolution or origin of life	Criteria of Category IV + prohibition of destructive impact or leakages after Earth-return, total containment of hardware directly in contact with extraterrestrial materials, absolute quarantine of extraterrestrial materials, continuous monitoring of returned hardware for replicating biochemical materials	Very high

Most of the Earth-orbiting satellites, which are part of the global Earth-observation and communications network, are slotted in Category I. However, missions bound for the Moon, near-Earth asteroids and Mars could be slotted anywhere between Categories II and V. As interplanetary connectivity becomes the centrepiece of the global space economy, the space-based assets and infrastructure will have to abide by the planetary protection norms, particularly when they lodge on the South Pole of the Moon and on Mars.

Another critical concern will be to prevent forward biochemical and organic contamination on extraterrestrial bodies and high-traffic interplanetary trajectories, i.e., transportation paths. These bodies and trajectories are critical from the purview of science. They might be critical for highly specialised inquiries about the presence of evidence of the origin of life on Earth. However, the same quest is vital from a security standpoint. If present, any disregard for extraterrestrial biological and chemicals will lead to backward contamination, leading to severe scenarios, including pandemics.

The current knowledge about the survivability limits of Earthly biological life in outer space's extreme environments is miniscule. Viruses and prions (strands of proteins that may assume infectious forms) are considered extreme forms of life in the universe. In the past, space missions to Mars, Moon, near-Earth asteroids, and comets have found that precursors to proteins and their basic building blocks are amino acids. The most restrictive Category V will apply to interplanetary infrastructure components returning to Earth in such a scenario. They will need thorough quarantine and inspection for any traces of backward contamination with viruses, prions and mutated microorganisms of Earthly origins.

Long-duration human presence in outer space will demand extensive quarantine facilities, protocols, prohibitions and necessary personnel to monitor and maintain the entire process. These biocontainment facilities will have to closely work with astronaut training, space medicine, life-support systems, space agriculture and other related domains.

Human access to the Moon, near-Earth asteroids and Mars is a recent occurrence. The many accomplished planetary exploration missions have reiterated that their milieu cannot provide the same ecological resonance with humans and other life forms as we have here on our native Earth. Nevertheless, many national leaders and the space agencies at their disposal have already explicated their willingness to continually study human biological compatibility and ergonomic efficiency operating in closed ecological life-support systems. The International Space Station, the ESA's Moon Village, the Russian Mars 500 experiment, the UAE's Mars Science City and the Indian modular space station are all part of these efforts. The copious amount of technical data these life-support platforms generate will help us to understand the impact of long-duration space travel, low gravity and space radiation environments on human biophysics and biochemistry. Therefore, it will take much more than health compatibility to make humans an interplanetary species.[6]

Even if substantial scientific progress affording humans the necessary biochemical compatibility towards becoming an interplanetary species, the world is not a cultural monolith. Once humans achieve the necessary biochemical compatibility, the next important task would be to analyse whether all civilisations, past and present, their respective repository of knowledge and populace could be made interplanetary.

Therefore, all spacefaring nations, regardless of their competence or astropolitical blocs they belong to, must comprehend their holistic aspirations for deep space exploration and the philosophies they seek to advance by attaining greater mobility and connectivity in interplanetary space. This realisation will be a significant task for humankind.

Notes

1 N. Bostrom, "Existential risks: Analyzing human extinction scenarios and related hazards," *Journal of Evolution and Technology* 9, (2002), https://ora.ox.ac.uk/objects/uuid:827452c3-fcba-41b8-86b0-407293e6617c/download_file?file_format=pdf&safe_filename=Existential%2Brisks%253A%2Banalyzing%2Bhuman%2Bextinction%2Bscenarios%2Band%2Brelated%2Bhazards&type_of_work=Journal+article

2 Retrieved from the *United Nations Office for Outer Space Affairs* website, https://www.unoosa.org/oosa/en/ourwork/spacelaw/treaties/outerspacetreaty.html

3 US Congress, "H.R.2617 – Apollo Lunar Landing Legacy Act," *13th US Congress* (2013–2014) https://www.congress.gov/bill/113th-congress/house-bill/2617

4 "Space2030: Space as a driver for peace," *United Nations Office for Outer Space Affairs*, 25 September 2018, https://www.unoosa.org/oosa/en/outreach/events/2018/spacetrust.html

5 "COSPAR policy on planetary protection," *Committee on Space Research – Panel on Planetary Protection*, 2020, https://cosparhq.cnes.fr/assets/uploads/2020/07/PPPolicyJune-2020_Final_Web.pdf

6 R.B. Setlow, "The hazards of space travel," *EMBO Reports* 4, (2003), 1013–1016.

8 Conclusion

The modern world is transitioning from the First to the Second Space Age. Many nations' space ambitions define these transitions and are no more limited to the astropolitical bipolarity between the US and the Soviet Union. In the Second Space Age, many nations, inconsequential during the First Space Age, are pursuing myriad space exploration capabilities. This advancing cohort is not delimiting space exploration activities to governmental space agencies but is aggressively rearing private innovation, manufacturing and service-driven industries. Some of these are employing their core competence in advanced Industry 4.0 sectors for space exploration activities. Economic and military superpowers will lead the astropolitical blocs in the Second Space Age, with each bloc pursuing its metastrategic goals. However, the bloc leaders realise their inability to achieve their objectives without partnerships and consensus with new spacefaring nations.

Those party to an astropolitical bloc are likely to mutually invest in interplanetary connectivity infrastructure, including transportation, communications, habitat, resource extraction and logistics networks between Earth, near-Earth asteroids, Moon and Mars, thus, giving them collective access to these crucial celestial bodies of the inner Solar System. The infrastructure will not overwhelm the 1967 United Nations Outer Space Treaty and its follow-up Rescue Agreement of 1967, the Space Liability Convention of 1972 and the Registration Convention of 1976. However, while constructing this interplanetary infrastructure, these countries can exploit the limitations in these treaties, pushing multilateral organisations to set up newer norms, regulations and codes of conduct. These astropolitical blocs will acquire the most considerable stakes possible in the global space economy that will grow by nearly 2.5 trillion dollars in the next two decades. Their race for the most significant stakes will cause the manipulation of loopholes in the existing space conventions.

DOI: 10.4324/9781003152934-8

Increased space activities will also create intense scrutiny of the nationalism versus globalism debate. Concepts like home, citizenship, statelessness, identity, crime, human rights, rescue, refuge, medical security, business damages, contract laws will naturally extend to the interplanetary infrastructure. An amplified extraterrestrial presence of citizens of various nations and their numerous space-based economic activities will also stimulate debates on ethics, business regulation, the scope of national security, terrorism and war. Furthermore, questions on the territorial goals of astropolitical blocs, the future nature of work for their human resource, on artificial and technology-driven alteration of the human body to enhance physiological and psychological capabilities, and philosophical questions about humanity's collective and trans-civilisational purpose in the universe will come under intense deliberation.

Astropolitical blocs will undoubtedly allow access and services to their infrastructure only to nations friendly to them. Raising this infrastructure will begin to blur lines between the private and public sectors to develop more remarkable intertwining. Such public–private collaborations are coming forth in many countries, including India, that have long avoided large-scale private sector participation. Some like New Zealand have never possessed a dedicated space agency until recently.

As more nations pursue their national space programmes for commercial, governmental and military gains, they will feel a need for outer space equivalent to the United Nations Convention on the Law of Sea (UNCLOS). Endurability and habitability are vital attributes that help seafarers survive on high seas. Ashore is never far away from any ship travelling along sea lanes and quick rescue is possible primarily. But, none of these attributes apply to the enormous, biologically and mechanically hostile, and inaccessible expanses of outer space. This demand will expand the scope of the existing Rescue Agreement and establish the Convention on the Law of Interplanetary Space that regulates all economic, transportation, connectivity, habitat activities in interplanetary space. Separately, a celestial body-specific – Convention of the Law of the Moon, Convention of the Law of near-Earth Asteroids and Convention of the Law of Mars – legal mechanisms for extraterrestrial surfaces will also become indispensable.

With amplified space economic activities, many space companies will feel the need to go public for their finances, giving rise to a securities asset class such as equities, fixed incomes, commodities and infrastructure as an asset class. These financial instruments will provide the necessary fillip to their national space economies and reduce overdependency on government budget allocations for national space programmes.

The next three decades are hugely important as the world enters into the Fourth Industrial Age and in the Second Space Age. These two phenomena will demand the breaking down of old notions of outer space activities and forming new lucrative mechanisms. However, to sustain profits, it will be essential to set rules for the game. The natural hostility of the space environment does not give enough scope for human conflict. Hence, cooperative competition between astropolitical blocs will be as important as competitive cooperation within them.

India's role in interplanetary connectivity will be pivotal. As a newly industrialised third-largest economy of the world, with a massive young population, an immense innovation potential, an enormous manufacturing prowess and a robust domestic market operating under transparent democratic norms, India will be a vital bridge between the developed and developing world aspirations from interplanetary connectivity.

Healthy socio-economic, military, scientific and environmental indicators, decisive leadership, pragmatic international partnerships, vigilant security apparatus and furtherance of the doctrine of strategic autonomy will be decisive for India's stakes in constructing and upkeeping interplanetary connectivity. In all this, it will be vital for New Delhi to comprehend that only through a robust Indian stake in the global space economy will it augment and sustain healthy socio-economic growth.

India will also have to shed its reticence for avante-garde space activities. Going a step ahead, it must secure its space technology innovation, manufacturing and services ecosystems from hostile investments, mergers and acquisitions. And therefore, it will need to raise its vigil in innovation, manufacturing and services sectors related to the space economy. Overall, India's participation in the evolving space economy will likely depend on the furtherance of the doctrine of strategic autonomy in outer space. It should partner with all the astropolitical blocs but contingent on its metastrategic ambitions.

Index

4 032